SEA BA

ENSURING JOINT FORCE ACC… …THE SEA

Committee on Sea Basing: Ensuring Joint Force Access from the Sea
Naval Studies Board
Division on Engineering and Physical Sciences

NATIONAL RESEARCH COUNCIL
OF THE NATIONAL ACADEMIES

THE NATIONAL ACADEMIES PRESS
Washington, D.C.
www.nap.edu

THE NATIONAL ACADEMIES PRESS 500 Fifth Street, N.W. Washington, DC 20001

NOTICE: The project that is the subject of this report was approved by the Governing Board of the National Research Council, whose members are drawn from the councils of the National Academy of Sciences, the National Academy of Engineering, and the Institute of Medicine. The members of the committee responsible for the report were chosen for their special competences and with regard for appropriate balance.

This study was supported by Contract No. N00014-00-G-0230, DO #21, between the National Academy of Sciences and the Department of the Navy. Any opinions, findings, conclusions, or recommendations expressed in this publication are those of the author(s) and do not necessarily reflect the views of the organizations or agencies that provided support for the project.

International Standard Book Number 0-309-09517-4

Additional copies of this report are available from:

Naval Studies Board, National Research Council, The Keck Center of the National Academies, 500 Fifth Street, N.W., Room WS904, Washington, DC 20001; and

The National Academies Press, 500 Fifth Street, N.W., Lockbox 285, Washington, DC 20055; (800) 624-6242 or (202) 334-3313 (in the Washington metropolitan area); Internet, http://www.nap.edu.

Printed in the United States of America

THE NATIONAL ACADEMIES

Advisers to the Nation on Science, Engineering, and Medicine

The **National Academy of Sciences** is a private, nonprofit, self-perpetuating society of distinguished scholars engaged in scientific and engineering research, dedicated to the furtherance of science and technology and to their use for the general welfare. Upon the authority of the charter granted to it by the Congress in 1863, the Academy has a mandate that requires it to advise the federal government on scientific and technical matters. Dr. Ralph J. Cicerone is president of the National Academy of Sciences.

The **National Academy of Engineering** was established in 1964, under the charter of the National Academy of Sciences, as a parallel organization of outstanding engineers. It is autonomous in its administration and in the selection of its members, sharing with the National Academy of Sciences the responsibility for advising the federal government. The National Academy of Engineering also sponsors engineering programs aimed at meeting national needs, encourages education and research, and recognizes the superior achievements of engineers. Dr. Wm. A. Wulf is president of the National Academy of Engineering.

The **Institute of Medicine** was established in 1970 by the National Academy of Sciences to secure the services of eminent members of appropriate professions in the examination of policy matters pertaining to the health of the public. The Institute acts under the responsibility given to the National Academy of Sciences by its congressional charter to be an adviser to the federal government and, upon its own initiative, to identify issues of medical care, research, and education. Dr. Harvey V. Fineberg is president of the Institute of Medicine.

The **National Research Council** was organized by the National Academy of Sciences in 1916 to associate the broad community of science and technology with the Academy's purposes of furthering knowledge and advising the federal government. Functioning in accordance with general policies determined by the Academy, the Council has become the principal operating agency of both the National Academy of Sciences and the National Academy of Engineering in providing services to the government, the public, and the scientific and engineering communities. The Council is administered jointly by both Academies and the Institute of Medicine. Dr. Ralph J. Cicerone and Dr. Wm. A. Wulf are chair and vice chair, respectively, of the National Research Council.

www.national-academies.org

COMMITTEE ON SEA BASING: ENSURING JOINT FORCE ACCESS FROM THE SEA

HARRY W. JENKINS, JR., ITT Industries, *Co-chair*
RICHARD L. WADE, Exponent, *Co-chair*
JEFFERY P. BENNETT, LMI
ALAN BERMAN, Applied Research Laboratory, Pennsylvania State University
PAUL BEVILAQUA, Lockheed Martin Aeronautics Company
E. RICHARD DIAMOND, JR., Raytheon Integrated Defense Systems
STEVEN W. FLOHR, Madison, Alabama
WESLEY L. HARRIS, Massachusetts Institute of Technology
GEORGE B. HARRISON, Georgia Tech Research Institute
KEVIN F. KELLY, Northrop Grumman Ship Systems
RONALD K. KISS, Webb Institute
JOHN B. LaPLANTE, Poland, Ohio
HENRY S. MARCUS, Massachusetts Institute of Technology
IRWIN MENDELSON, Delray Beach, Florida
WILLIAM B. MORGAN, Rockville, Maryland
JOHN H. MOXLEY III, Korn/Ferry International
ROBERT G. SPRIGG, General Dynamics/Bath Iron Works
JENNIFER P. WHITLOCK, The Boeing Company

Staff

CHARLES F. DRAPER, Naval Studies Board Acting Director (as of July 12, 2003)
ARUL MOZHI, Study Director
MARY G. GORDON, Information Officer
SUSAN G. CAMPBELL, Administrative Coordinator
IAN M. CAMERON, Research Associate
SEYMOUR J. DEITCHMAN, Consultant
SIDNEY G. REED, JR., Consultant
RAYMOND S. WIDMAYER, Consultant

Preface

In response to the evolving U.S. national security posture, the Navy and Marine Corps promulgated a naval vision entitled *Naval Power 21,* based in part on "assuring *sea based* [emphasis added] access worldwide for military operations, diplomatic interaction, and humanitarian relief efforts."[1] More importantly, Sea Basing "projects the sovereignty of the United States globally while providing joint force commanders with vital command and control, fire support, and logistics from the sea, thereby minimizing vulnerable assets ashore."[2] Thus, Sea Basing represents a force-entry concept with potential for U.S. and coalition military forces in various regions of the world, thereby affecting the strategic landscape from which these forces will ultimately project power.

Recent events in Kosovo, Afghanistan, and Iraq have underlined the fact that the availability of land bases during conflicts may be uncertain owing to physical or political factors that delay, limit, or prevent their use. In many future military situations, the assumption of readily available, secure land bases is likely to be open to question, although one can make the argument that the Naval Services operate from a sea base of sorts today.

---

[1]Hon. Gordon England, Secretary of the Navy; ADM Vern Clark, USN, Chief of Naval Operations; and Gen James L. Jones, USMC, Commandant of the Marine Corps. 2002. *Naval Power 21,* Department of Defense, Washington, D.C., p. 1.

[2]Hon. Gordon England, Secretary of the Navy; ADM Vern Clark, USN, Chief of Naval Operations; and Gen James L. Jones, USMC, Commandant of the Marine Corps. 2002. *Naval Power 21,* Department of Defense, Washington, D.C., p. 5.

Specifically, Carrier Battle Groups (CVBGs) carry a certain amount of organic sustainment that supports their power-projection mission. Amphibious Ready Groups (ARGs) carry 10 to 15 days' worth of all classes of supplies that sustain initial operations by Marines ashore. Both the CVBGs and the ARGs, or the new Carrier Strike Groups and Expeditionary Strike Groups as recently promulgated by the Chief of Naval Operations and the Commandant of the Marine Corps,[3] are reinforced at sea (if required) by the Navy's combat logistics force ships and the three Maritime Prepositioning Force (MPF) squadrons. The MPF squadrons, each carrying a 30-day supply of equipment for a 16,000-person brigade force, can support operations ashore if larger Marine Air Ground Task Forces, at the Marine Expeditionary Brigade and Marine Expeditionary Force levels, are required to support larger assigned missions. Unfortunately, it is difficult to off-load the three MPF squadrons except in port, and therefore they have minimal in-stream logistics capability otherwise when deployed.

Connectors that enable Sea Basing are an issue related to its implementation. For example, the connectors to move forces from facilities in the continental United States (CONUS) to advanced bases so that forces could eventually marry up with material, and ultimately to move forward forces and material to military objectives, are a crucial facet of Sea Basing. Recently, the Navy and the Army have experimented with high-speed ferry ships that have some capabilities as connectors.

Most recently, the Defense Science Board (DSB) Task Force on Sea Basing concluded that Sea Basing will be a critical future joint military capability and that it will help ensure access to areas in which U.S. military forces might be denied access to support facilities.[4] In addition, the DSB concluded that it will be essential for the Department of Defense (DOD) to undertake a spiral development approach—one that accounts for evolution from current capabilities to tomorrow's needs. In recommending such an approach, the DSB identified 12 issues for the realization of Sea Basing: 6 relate to management, planning, and resources, and the remaining 6 are grouped as new capabilities, some of which include issues relating to the connectors needed for enabling Sea Basing. Among the 6 new capabilities, the DSB identified 3 as important needs for the implementation of Sea Basing: (1) the capability to handle cargo in rough seas, characteristic of many likely areas of operations; (2) a heavy-lift aircraft (able to transport more than 20 tons) with theater-wide range that can be based at sea; and (3) ships whose design incorporates all of the requirements of the sea base system of systems.

---

[3]ADM Vern Clark, USN, Chief of Naval Operations; and Gen Michael W. Hagee, USMC, Commandant of the Marine Corps. 2003. *Naval Operating Concept for Joint Operations,* Department of Defense, Washington, D.C., September 22, p. 7.

[4]Defense Science Board. 2003. *Defense Science Board Task Force on Sea Basing,* Office of the Under Secretary of Defense for Acquisition, Technology, and Logistics, Washington, D.C., August.

In April 2004, the Department of the Navy requested that the National Research Council (NRC), under the auspices of the Naval Studies Board, convene a workshop to assess the science and technology base, both inside and outside the Department of the Navy, for developing Sea Basing, and that it identify research priorities for supporting future Sea Basing concepts. The terms of reference of the workshop are provided below. MajGen Harry W. Jenkins, Jr., USMC (retired), of ITT Industries, and Richard L. Wade of Exponent co-chaired the Committee on Sea Basing: Ensuring Joint Force Access from the Sea. Biographical information on the committee membership and the NRC staff is presented in Appendix A.

The purpose of the committee's brief effort differs from that of the DSB's more extensive study. The present report can help the Department of the Navy focus better on what it needs to do, whereas the broader DSB effort addresses the needs of the entire Department of Defense. The committee did not intend to enlarge upon, augment, or replace the DSB work, but rather to form some independent judgments, based on the experience and backgrounds that the committee members brought to the workshop that it held in September 2004 (see Appendix B) and on the information that was given to the committee by the Services and the DOD. Also, in the limited amount of time available for this review, it was not possible to study aircraft and ship design problems in any depth. Thus, it was necessary to rely heavily on the judgments of the workshop participants, which were based on their prior knowledge of the ship and aircraft design issues and the possible solutions to them.

TERMS OF REFERENCE

At the request of the Department of the Navy, the Naval Studies Board of the National Research Council convened a workshop to assess the science and technology base, both inside and outside the Department of the Navy, for developing Sea Basing, and to identify research priorities for supporting future Sea Basing concepts. Specifically, the workshop addressed the following:[5]

- Current and future naval operational concepts related to Sea Basing, including plans for, opportunities from, and limitations of Sea Basing; and
- A technology roadmap with anticipated time horizons for the realization of the following: (1) the capability to handle cargo in heavy sea states; (2) a long-range, heavy-lift aircraft (able to transport more than 20 tons) based at sea; and (3) ship classes and warfare systems that support open architecture design considerations.

---

[5]Because of the limited nature of this effort, a number of topics could not be addressed: for example, the Mobile Offshore Base, modularity of logistics units, medical logistics, dependency on environmental predictions, and tactical trade-offs to accommodate logistics results.

In addition, the committee was tasked to assess the feasibility of the Naval Studies Board's conducting a future study on Sea Basing aimed at identifying the Department of the Navy's broader needs for developing a Sea Basing capability to support naval (and joint) force operations.[6]

STUDY WORKSHOP AND REPORT PREPARATION

The committee conducted a 4-day workshop—September 7-10, 2004—at the Keck Center of the National Academies in Washington, D.C. (see Appendix B for the workshop agenda). During the workshop, the committee heard presentations from the Department of the Navy, the U.S. Army, the Office of the Secretary of Defense, and industry. The committee then outlined an initial draft report.

The months between the workshop and the publication of this report were spent preparing the draft manuscript, gathering additional information, reviewing and responding to external review comments, editing the report, and conducting the required security review to produce a public report.

[6]To address this task, the committee identified a number of studies that should be performed in a joint context in order to fix the proposed design and operating parameters of the sea base and airborne and seaborne connectors.

Acknowledgment of Reviewers

This report has been reviewed in draft form by individuals chosen for their diverse perspectives and technical expertise, in accordance with procedures approved by the National Research Council's (NRC's) Report Review Committee. The purpose of this independent review is to provide candid and critical comments that will assist the institution in making its published report as sound as possible and to ensure that the report meets institutional standards for objectivity, evidence, and responsiveness to the study charge. The review comments and draft manuscript remain confidential to protect the integrity of the deliberative process. We wish to thank the following individuals for their review of this report:

Roy R. Buehler, Mableton, Georgia,
Roy C. Evans, Jr., The MITRE Corporation,
Edward A. Frieman, Scripps Institute of Oceanography,
Peter A. Gale, J.J. McMullen Associates,
Michael P. Kalleres, Jacksonville, Florida,
J. Randolph Paulling, Geyserville, California,
Edward Petrushka, Fort Worth, Texas, and
John E. Rhodes, Balboa, California.

Although the reviewers listed above provided many constructive comments and suggestions, they were not asked to endorse the conclusions or recommendations, nor did they see the final draft of the report before its release. The review of this report was overseen by William G. Howard, Jr., Scottsdale, Arizona.

Appointed by the National Research Council, he was responsible for making certain that an independent examination of this report was carried out in accordance with institutional procedures and that all review comments were carefully considered. Responsibility for the final content of this report rests entirely with the authoring committee and the institution.

Contents

Executive Summary

The sea base now under discussion is viewed as an enabler of joint force campaigns. It is conceived as a means of projecting joint power and influence ashore, without the need for large port and airfield terminals that are physically vulnerable to attack and that may be subject to sovereign constraints by the nations in which the bases are embedded. The sea base, protected by the Navy's Sea Shield,[1] would provide the capability for joint forces to enter an area, forcibly if necessary, and move rapidly against the main objective while sustaining themselves from the sea, either until they could establish secure ports and airfields ashore or for the entire duration of the operation.

The number, strength, and size of units of action that could be sustained from a sea base and the time over which they could be supported remain to be determined. Early discussions indicate the joint Service desire that the sea base be designed to support, among other things, a joint command center afloat, Special Operations Forces ashore, and logistic support of any forces that are ashore up to the design level (yet to be determined) of the sea base. The naval forces have expressed a need to land and sustain a force up to and including a Marine Expeditionary Brigade. The Army has discussed landing and sustaining one or more

[1]Sea Shield will provide a layered defense to protect the homeland, sustain access to contested littorals, and project a defensive umbrella over coalition partners and joint forces ashore in distant theaters. See VADM Michael Bucchi, USN; and VADM Michael Mullen, USN, 2002, "Sea Power 21 Series, Part II: Sea Shield: Projecting Global Defensive Assurance," *U.S. Naval Institute Proceedings*, Vol. 128, No. 11. Available online at <http://www.usni.org/Proceedings/Articles02/PRObucchi11.htm>. Last accessed June 2005.

brigade combat teams, whose size and military strength remain to be determined. Long-range Sea Strike[2] naval fires, in addition to Marine, Navy, and Air Force air-combat capability, will support Army forces, naval forces, and Special Operations Forces ashore. Theater-support airlift, with precision airdrop capabilities if needed, will also be available to sustain the forces ashore as appropriate and feasible.

The brief review conducted by the Committee on Sea Basing: Ensuring Joint Force Access from the Sea and reported here examined the state of planning for the sea base, including the following:

- The concepts of operation (Chapter 1);
- The state of the ship and aircraft technology available to make the sea base work, including the difficult problem of cargo transfer in sea states up to and including Sea State 4 (Chapters 2 through 4); and
- The issues involved in creating the sea base as a joint system of systems (Chapter 5).

KEY CONCLUSIONS

As a result of the workshop presentations and its own deliberations, the committee reached the following key conclusions:

1. Planning for the sea base is still in its infancy. Coordination across the Services is just beginning, with discussions between the Navy, the Army, and the Marine Corps, and in response to the Office of the Secretary of Defense (OSD) initiatives.[3] As a consequence, the details have yet to emerge, especially with respect to joint operations and with respect to the availability and needed development of technology to perform the tasks that are envisioned for the sea base. As an example, there is just beginning to be planning that reflects the implications of the heavy-lift aircraft or the Army's shallow draft high-speed ship (SDHSS) concept for the design of a sea base—or, on the other hand, the implications of a sea base for the heavy-lift aircraft or the SDHSS. A summary of the committee's findings and conclusions regarding current and future naval operational concepts related to Sea Basing—including plans for, opportunities from, and limitations of Sea Basing—is given in Box ES.1.

---

[2]Sea Strike is a broad concept for naval power projection that leverages command, control, communications, computers, combat systems, intelligence, surveillance, and reconnaissance; precision; stealth; information; and joint strike together. See VADM Cutler Dawson, USN; and VADM John Nathman, USN, 2002, "Sea Power 21 Series, Part III: Sea Strike: Projecting Persistent, Responsive, and Precise Power," *U.S. Naval Institute Proceedings*, Vol. 128, No. 12. Available online at <http://www.usni.org/Proceedings/Articles02/PROdawson12.htm>. Last accessed June 2005.

[3]These initiatives include the Joint Vertical Aircraft Task Force and the Army-led Team on Heavy-Lift Aircraft for Seabasing.

BOX ES.1
Summary of Findings and Conclusions Regarding Current and Future Naval Operational Concepts Related to Sea Basing

Findings

• *Plans for Sea Basing.* Sea Basing concept documents, both Service and joint, have been issued or are in draft. While they have much in common, there are significant differences among them. Tactical and programmatic planning is underway, but it lacks a unified vision and coordination among the myriad elements, and although discussions between the Army and Marine Corps have begun, the approach does not yet involve joint entities outside the Department of the Navy in any significant way.

• *Opportunities from Sea Basing.* Sea Basing has the potential to change—in very fundamental respects—the way that the armed forces of the United States operate. Among the most important potential benefits of Sea Basing are these: assurance of operational access, enhanced forward-defense posture, profound improvement in immediate response capability, rapid initiation of joint command and control (C2), very rapid transition from crisis to joint forcible entry, and an unprecedented degree of force tailorability and scalability.

• *Limitations of Sea Basing.* Operational Sea Basing is potentially limited by its dependence on the capabilities resident in Sea Shield. It will require robust command and control—of fires, maneuver, logistics, and operations—at great distance. The C2 systems and communications to support this function are limited, and development efforts are uncoordinated and desultory. There will also be sea-state limitations, space limitations on the sea base, and limitations imposed by the need to transfer cargo between the connectors and the sea base in the open ocean in a seaway. Sea Basing has obvious benefits in conducting limited regional conflicts, but it appears to have a more limited or complementary role in major regional conflict or theater warfare, because of the limitations in the flow of supply tonnage, most of which would have to arrive by sea, that could be sustained in such a situation.

• *End state.* The Department of the Navy has sent mixed messages with regard to the level of Sea Basing capability desired. It is not clear if the sea base is to be capable of projecting full military power, as was concluded by the Defense Science Board, or is to be something more modest.

Conclusions

• The sea base should be configured to the desired end state of full joint integration, permitting the employment and sustainment of a wide range of joint forces from the sea base. In this configuration, the sea base is fully joint rather than a purely naval capability.

• Procurement programs that would be required at any ultimate level of sea base capability need to be emphasized. These include high-speed connectors; landing craft air cushion replacement; heavy-lift, intratheater airlift; Sea State 4 operating capability; Maritime Prepositioning Force (Future) design; and a strike-up/strike-down system.

continued

BOX ES.1 Continued

- Sea Basing programs and development efforts need to be brought under common joint management through the establishment of a Joint Sea Base Planning Office, under the Office of Secretary of Defense. The responsibilities of this office would include procurement, doctrine development, war gaming, and any other activity with the potential to effect the attainment of this transformational capability.
- The proposed Joint Sea Base Planning Office should take immediate steps to involve the other Services, the combatant commanders, and appropriate defense agencies in the development of the Sea Basing capability.

2. Some critical technology developments on which the success of the sea base will depend have yet to be realized. They include the following:

a. Heavy-lift aircraft that can carry a standard International Organization for Standardization (ISO) 20 ft container[4] or its full contents, depending on the ability of troops ashore to handle the loading and off-loading, or that can carry the Army's Stryker combat vehicle, and that can operate from a main ship of the sea base and carry such loads from the sea base to an operational radius of 150 to 300 nautical miles (nmi). Several possibilities are being considered, including rotary and fixed-wing, vertical takeoff and landing (VTOL), and short takeoff and landing (STOL). The committee believes that obtaining the desired combination of payload and range could lead to a fixed-wing aircraft having powered lift of some as-yet-undetermined configuration, with the approximate payload capacity of a C-130J, and that the aircraft will be able to operate in super-short-takeoff-and-landing (SSTOL) or short-takeoff-and-vertical-landing (STOVL) mode, and possibly in full VTOL mode;

b. Cargo-transfer capability from the main ships of the sea base to seaborne connector ships in sea states as high as Sea State 4; and

c. One or more surface ships designed to operate the airborne connectors and to interface with the various seaborne types of connectors between the advanced base, the sea base, and objective areas or intermediate landing zones on a beach. Incorporating in a single ship the capability for ship-to-ship cargo transfer and a flight deck large and strong enough to handle aircraft of the size

[4]These freight containers conform to the ISO container-manufacturing standards. The committee assumed an ISO standard 20 ft × 8 ft × 8.5 ft container, loaded so that the weight of the container plus payload is not over 40,000 lb for the "threshold" value and not over 50,000 lb for the "objective" value. The committee did not discuss the feasibility of smaller supply or vehicular modular units (except for the assembly and disassembly of the CH-53 helicopter).

and weight implied in item 2a (above) will present especially demanding design and operational problems, which will most likely lead to the need for two different types of ships in the sea base. In either case, cargo handling in Sea State 4 and operations with heavy-lift aircraft in the 20 to 23 ton payload weight class will lead to very large ships.[5]

3. The typical design, development, and acquisition times for the new ship(s), for the ship-to-ship cargo-transfer capabilities in high sea states, and for the heavy-lift aircraft systems will, judging from experience, be different. They will need to be reconciled in order to establish schedules for appropriately phased acquisition and for initial and full operational capability for a sea base as a whole. Of the key technology drivers, the expected times of operational capability for a heavy-lift aircraft of new design will be in the decade of the 2020s—possibly in 2015 for the Marine Corps CH-53(X) if it is started in 2006, and later than 2025 for more-advanced heavy-lift aircraft such as quad-tilt rotor, tilt-wing, or fan-in-wing configurations in the 20 to 23 ton payload class. Capabilities for ship-to-ship cargo transfer in high sea states can be expected to become available in the period from 2012 to 2022, depending on the extent to which they depart from the current state of practice. It will thus be necessary to plan the development and deployment of a sea base in phases to accommodate such differences.

4. Among the constraining operational problems will be those emerging from the need for simultaneous airborne and seaborne cargo transfer from the sea base in order to meet the needs of forces ashore. Seaborne cargo-transfer operations will require a ship heading that protects the transfer from the adverse effects of wind and waves; airborne transfer operations, especially if the heavy-lift aircraft is a fixed-wing aircraft needing some takeoff run, will require a ship to head into the wind at significant speed. Thus, for simultaneous airborne and seaborne cargo transfer, an intricate choreography of cargo sequencing from single ships, or coordinated operations for multiple ships, will be required, depending on whether the sea base includes single ship types capable of air operations and cargo transfer to other ships, or two types of ships for the two distinctly different kinds of cargo-transfer operations. These considerations will become design issues affecting the overall size and cost of the sea base, or else they will become limiting factors in the resupply rates that sea bases of a given size and capacity will be able to achieve or in the size of the forces that they will be able to support.

5. The sea base must emerge from a joint Service effort. The successful development of joint Sea Basing capabilities for the nation will involve long-term

[5]The committee was not briefed on and did not discuss the Mobile Offshore Base concept. Several recent studies, including the 2003 Defense Science Board report on Sea Basing, discuss the advantages and disadvantages of this concept. See Defense Science Board, 2003, *Defense Science Board Task Force on Sea Basing,* Office of the Under Secretary of Defense for Acquisition, Technology, and Logistics, Washington, D.C., August.

capital investments in interdependent, complex platforms, connectors, and supporting technologies, some of which exist today or are under development and some of which are yet to be developed. To succeed, such an effort must of necessity involve all military Services and appropriate U.S. government agencies in a coordinated development process under a coherent set of ultimate goals, common standards, integrated requirements, and time-phased, mutually accepted priorities. It is unlikely that in the current budget climate there will be resources for different Services' variously designed major ships, connector ships, and heavy-lift aircraft for the sea base. For these reasons, all elements of the sea base must have common ships and cargo-transfer components across the Services and Special Operations Command, all fully compatible with the Service cargos to be moved. By virtue of their experience and capability in ship design, shipbuilding, and at-sea operations, the Navy and Marine Corps should be designated lead Services for this joint effort.

6. The development of the sea base will require a major ship as a testbed to support the necessary experimentation, including that with cargo transfer at sea, which will determine the ultimate sea base configuration and operating procedures. Various ships could serve this purpose in part: an existing large commercial ship; a large, medium-speed roll on/roll off (LMSR) or existing Maritime Prepositioning Force (MPF)-type reserve ship in a reduced operating status, such as an SL-7/T-AKR Fast Sealift Ship, or some other comparable ship. Experiments with the heavy-lift aircraft will require extensive modifications of the superstructure of such a ship and the construction of a flight deck strong enough to take the loads of very large aircraft. If the modification of an existing large ship proves technically infeasible for the purpose, at-sea experiments with fixed-wing aircraft in the 20 to 23 ton payload weight class may have to be performed by activating a reserve carrier or awaiting the design and construction of the lead ship of a new ship class designed for the purpose.

7. A number of investigations, which should be performed in a joint context, are required to fix the proposed design and operating parameters of the sea base and airborne and seaborne connectors. The design parameters of the sea base—including the forces that it should be designed to support, its operating distance from shore, and the kinds of ships that will be needed—cannot be finally decided until the results of the first three of the following studies are available. The required studies are as follows:

a. A technical systems study—involving the trade-offs, interactions, and costs among the various possible heavy-lift aircraft types, the major ships of the sea base, and the seaborne connector ships—is needed to settle on the preferred design parameters of the total sea base system. Trade-off studies are also required for the alternative methods of effecting cargo transfers between the sea base and the connectors—both the supply connectors and the shore-bound connectors. The selected transfer methods can drive the designs of both the sea base ships and the connectors themselves. These studies should be coordinated with the Air Force's AMC-X[6] and M-X[7] studies, the Army's SDHSS study, the ongoing Army-Marine Corps Advanced Technology Demonstration efforts, and the OSD-mandated Joint Vertical Aircraft Task Force.

b. An operational systems study is needed to determine the size of Army, Marine, and Special Operations Forces that sea bases of specific sizes and costs can practically sustain and therefore the levels of engagement for which to design the sea base. This study should take into account the potential for air cargo delivery to forces ashore from a distant intermediate or advanced base by precision airdrop or, if possible, landings (but not refueling) at the forward positions ashore. The study should also probe the limits of sea base distance offshore and the total area dispersion of a multiship sea base for simultaneous air- and sea-cargo-transfer operations, to ascertain the limiting parameters of Sea Shield to protect the sea base and the heavy-lift aircraft range requirements[8] for operation off the sea base.

c. A ship survivability study is needed to examine the levels of passive protection between fully militarized levels and commercial standards that should be built in to the main and connector ships of the sea base, and whether the ships should have active defenses to engage cruise missiles or aircraft that leak through Sea Shield.

d. Development is needed of provisional Sea Basing standards of design, commonality of cargo-transfer components, and compatibility or interoperability among Service-unique components, to guide the design of the interfaces of the sea base and the airborne and seaborne connectors.

e. Provisional development of joint doctrines and tactics, techniques, and procedures is needed to guide the early experimentation and ultimately the design and development of the sea base.

KEY RECOMMENDATIONS

Recommendation: A Joint Sea Base Planning Office—directed by a Navy flag officer or a Marine Corps general officer—should be established. The director of

[6]New aircraft called the Advanced Mobility Concept, or AMC-X, with about the same cargo capacity as a C-130 but able to fly higher and faster, while operating from 2,000 ft runways. Also, the AMC-X would be stealthy.

[7]M-X, Materiel, Experimental; designators assigned to U.S. Air Force research and development projects. M-X numbers were assigned very early in a product's evolution, and thus many M-X numbers were subsequently canceled or closed out without producing anything more than a research report.

[8]Aeromedical evacuation was not briefed to or discussed by the committee, but that will also bring into consideration the range from the sea base to the nearest advanced medical care facility. See Naval Studies Board, National Research Council, 1999, *Naval Expeditionary Logistics*, National Academy Press, Washington, D.C., pp. 54-55.

this office should report to an appropriate official in the Office of the Secretary of Defense. The Joint Sea Base Planning Office should be staffed with representatives from the four Services, Special Operations Command, and U.S. Transportation Command. These representatives should be at the O-6 level or above. This office should be responsible for carrying out the necessary studies for the design and operation of the sea base and the airborne and seaborne connectors and for guiding the experimentation with one or more testbeds, as needed. At some appropriate stage of planning, as the studies and experimentation and related Service programs mature, this office could grow into a Joint Program Office for the Sea Base.

Recommendation: The Joint Sea Base Planning Office should create a joint master plan for technology development, based on an integrated system-of-systems engineering approach, for the Services to use in developing the components of the sea base within their assigned jurisdictions.

Recommendation: The Department of the Navy should identify one large vessel to be used as a testbed for resolving the known problems, including those related to connectors and internal cargo handling, involved in at-sea cargo transfer at Sea States 3 and 4, or two such vessels if required for an integral flight deck in order to explore issues associated with potential future heavy-lift aircraft. The Department of the Navy should pursue private industry proposals for the acquisition of larger commercial vessels for such testbeds.

Recommendation: The Joint Heavy-lift Aircraft Exploration of Concepts being coordinated within the Office of the Secretary of Defense should involve the U.S. Transportation Command in the process. This effort should be transferred to the Joint Sea Base Planning Office when this office is created.

Recommendation: In order to facilitate the management of multi-Service efforts through the use of common terminology and technologies as they create the various elements of the sea base and the airborne and seaborne connectors, and to focus the overall effort, the entire sea base effort within the Joint Sea Base Planning Office should be managed under a name—for example: "Joint Maritime Prepositioning and Sea Basing Force."

A number of related or subordinate conclusions and recommendations at a more detailed level are included in the chapters of this report.

Finally, in response to the question in the terms of reference for this study (see the Preface) about whether the Naval Studies Board should conduct a further study in greater depth on Sea Basing, the committee believes that such a study would not be advisable at this time. When the studies listed above in the seventh key conclusion are accomplished by the recommended Joint Sea Base Planning Office (or in some other forum suitable for such detailed analyses), the necessary further study of the issues will have been accomplished.

1

Review of Sea Basing Concepts

A number of Sea Basing concept documents—some Service-originated and some joint documents—have been issued or are in draft form. While they have much in common, there are significant differences among them. For the purpose of this study, the committee chose to base its analysis on the *Naval Operating Concept for Joint Operations*, which was issued by the Navy and Marine Corps Service chiefs jointly. This document defines Sea Basing as—

> the foundation from which offensive and defensive power are projected, making Sea Strike[1] and Sea Shield[2] realities. It describes the projection, sustainment and operational maneuver of sovereign, distributed and networked forces operating globally from the sea. Sea Basing will provide Joint Force Commanders with global command and control (C2) capability and extend integrated support to other services.[3]

---

[1]VADM Cutler Dawson, USN; and VADM John Nathman, USN. 2002. "Sea Power 21 Series, Part III: Sea Strike: Projecting Persistent, Responsive, and Precise Power," *U.S. Naval Institute Proceedings*, Vol. 128, No. 12. Available online at <http://www.usni.org/Proceedings/Articles02/PROdawson12.htm>. Last accessed June 2005.

[2]VADM Michael Bucchi, USN; and VADM Michael Mullen, USN. 2002. "Sea Power 21 Series, Part II: Sea Shield: Projecting Global Defensive Assurance," *U.S. Naval Institute Proceedings*, Vol. 128, No. 11. Available online at <http://www.usni.org/Proceedings/Articles02/PRObucchi11.htm>. Last accessed June 2005.

[3]ADM Vern Clark, USN, Chief of Naval Operations; and Gen Michael W. Hagee, USMC, Commandant of the Marine Corps. 2003. *Naval Operating Concept for Joint Operation,* Department of Defense, Washington, D.C., September 22, p. 3.

NAVAL OPERATING CONCEPT FOR JOINT OPERATIONS

The Naval Operating Concept (NOC) describes the near-, mid-, and far-term organizational structures that are to evolve the concepts to achieve joint operational capability over the full spectrum of conflict. It states as a far-term objective "an increased ability to protect, project, and support joint and multinational forces from the sea."[4] As is the case with many concept documents, however, it is silent on the particulars. Specifically, the NOC does not further define "joint and multinational forces" with respect to size, type of unit, or means of employment. This lack of specificity has left the members of this committee, as well as the community at large, without a full definition of the desired end-state capabilities for this concept. For example, the Defense Science Board has stated that "sea basing must become a truly joint concept with capabilities that allow for the projection of the full panoply of military power."[5] The uncertainty regarding the ultimate destination of this concept is a serious problem, as is shown later in this chapter.

The committee's analysis of the NOC, together with a review of related documents, can be summarized as follows:

- *Plans for Sea Basing*. Sea Basing is one of three warfighting concepts, or pillars, basic to the the Navy's capstone concept Sea Power 21 (the others are Sea Strike and Sea Shield).[6] It is the only one of the three that represents a new mission and is, in the view of the committee, the most transformational of the three pillars. These characteristics of Sea Basing have led to the engagement of many organizations in the planning and programming necessary to achieve this capability. The existing planning for Sea Basing is tactical (e.g., the development of concepts of operations, and war gaming to explore and define operational parameters: distances, times, logistical needs, and so on), as well as programmatic (major acquisition programs, associated research and development (R&D), and so on). It did not appear to the committee, however, that this broadly based planning is coordinated or that there is any central direction guiding it. As a consequence, the planning is to a degree incoherent. Further, it has not involved joint entities outside the Department of the Navy (such as combatant commanders (COCOMs), the Air Force, Army, Transportation Command (TRANSCOM), Defense Logistics Agency (DLA), and the like) to any significant degree. For example, U.S. Army forces are not configured or conceptually designed in a sea

[4]ADM Vern Clark, USN, Chief of Naval Operations; and Gen Michael W. Hagee, USMC, Commandant of the Marine Corps. 2003. *Naval Operating Concept for Joint Operations,* Department of Defense, Washington, D.C., September 22, p. 16.

[5]Defense Science Board. 2003. *Defense Science Board Task Force on Sea Basing*. Office of the Under Secretary of Defense for Acquisition, Technology, and Logistics, Washington, D.C., August.

[6]ADM Vern Clark, USN, Chief of Naval Operations. 2002. "Sea Power 21 Series, Part I: Projecting Decisive Joint Capabilities," *U.S. Naval Institute Proceedings*, Vol. 128, No. 10. Available online at <http://www.usni.org/Proceedings/Articles02/PROcno10.htm>. Last accessed June 2005.

base context, although the Army itself is beginning to move in that direction. The potential consequences of these problems with planning are disturbing. For example, absent Army input, Navy planners might assume an incorrect or inappropriate Army force mix. If sea base vessels and support systems were configured to deliver a Stryker Brigade and the Army desired to use or assumed delivery of an Air Assault Brigade, considerable mismatch would result. However, the Navy, Army, Marine Corps, and TRANSCOM are initiating discussions to begin some of the necessary coordination in planning.

- *Opportunities from Sea Basing*. Sea Basing has the potential to change—in very fundamental respects—the way that the armed forces of the United States operate. Among the most important potential benefits of Sea Basing are (1) assuring access worldwide for military operations, (2) enhanced forward-defense posture, (3) improvement in immediate response capability, (4) rapid initiation of joint command and control, (5) very rapid transition from crisis to joint forcible entry, and (6) a greater degree of force tailorability and scalability.
- *Limitations of Sea Basing*. The limitations of Sea Basing can be characterized as both operational and cultural:

—Operationally, Sea Basing is dependent on the capabilities projected to be resident in Sea Shield, particularly those associated with missile defense, antisubmarine warfare, and mine countermeasures. Although Sea Basing has obvious benefits in conducting limited regional conflicts, it appears to have only a complementary role in major regional conflict or theater warfare. This is because of the limitations in the flow of supply tonnage—most of which would have to arrive by sea—that could be sustained in such a situation. Sea Basing will require robust command and control—of operations, fires, movement, and logistics. The command, control, and communications (C3) systems necessary to command and support tactical operations from the sea at great ranges (variously given as 110 to 300 nautical miles) are limited, and current development efforts are desultory at best.

—The practical physical limitations of Sea Basing include those caused by high sea states, restricted space on the sea base, and the limitations imposed by the need to transfer cargo between the connectors and the sea base in the open ocean in a seaway.

—The cultural limitations, however, are the more significant. To field a Sea Basing capability such as that described in the NOC will require realignment in procurement accounts, together with far-reaching changes in training, doctrine, force structure, and infrastructure. To achieve the Sea Basing capability, all of these changes will be necessary, and because of the nature of the needed changes, all will most likely be resisted vigorously.

—Finally, to date there is a lack of an overall integrated systems engineering approach, which is required to consider the multiple elements and increments necessary to yield the optimum Sea Basing capability. A system-of-systems approach is necessary in order to gain the synergy needed to achieve a successful joint Sea Basing capability.

IMPLEMENTATION OF SEA BASING CAPABILITIES

The extent to which Sea Basing is initially designed, developed, and implemented will determine the extent to which it achieves the envisioned joint operational capabilities. If Sea Basing is to be successful as a joint capability, it must be joint from its conception on. If fully implemented, the Sea Basing concept offers a significant new capability to protect, project, support, and sustain joint forces. If it is implemented to some lesser degree, the concept will fall well short of the envisioned attributes quoted at the beginning of this chapter.

The level of implementation has profound implications for the types of vessels that will comprise the sea base. The most critical of the questions regarding the level of implementation involve the aviation capability and the ability to move cargo between ships in sea states as high as Sea State 4, but the effect of the ultimate capability desired will reverberate throughout the entire design process.

The dependence of a future sea base on the systems embedded in Sea Shield is unsurprising, being consonant with historical practice. The Navy-Marine Corps doctrine governing amphibious assault requires air and sea superiority before an assault is launched. This doctrine must be brought up to date to encompass relatively new tactics such as over-the-horizon assault, possibly by substituting a form of tiered protection for an absolute requirement for superiority. However, the principle underlying any new doctrine will be the same as the historical principle: an assault force must rely on external protection for its survivability.

It is yet to be determined if commercial standards[7] for damage stability and many other factors such as strength are generally acceptable for the ships, the connector vessels, and other craft comprising a future sea base. Some upgrading of these standards in such areas as electrical redundancy and firefighting will be required. The sea base will operate alongside commercial container ships; commercial carriers for petroleum, oil, and lubricants; and allied and coalition vessels of all types: if the Sea Shield umbrella is not sufficient to provide a low degree of vulnerability for a force such as this, in all likelihood the Sea Basing operation would not be undertaken. To validate these conclusions and to quantify the issues, a vulnerability study is recommended.

The committee identified four levels of implementation of Sea Basing capability that could be pursued in the context of the Naval Operating Concept:

- *Level One—reinforcement and/or benign operations.* Level One is the least-expensive and lowest capability level. It represents an improvement over current capability and would entail the following: the reinforcement of an ongoing Expeditionary Strike Group (ESG)/Marine Expeditionary Unit (MEU) operation by a Marine Expeditionary Brigade (MEB) through use of the prepositioning of

[7]These commerical standards include those of the American Bureau of Shipping, the Maritime Administration, and the U.S. Coast Guard.

ships for providing equipment and sustainment support and strategic airlift for personnel delivery. The development of Sea Basing capabilities such as selective off-load, at-sea assembly, and improved delivery and transfer systems would greatly enhance the existing capability.

- *Level Two—MEB Ship-to-Objective Maneuver (STOM).* The Level Two option would enable the combatant commander to employ an MEB (or potentially a larger Marine Corps force) tactically from the sea. In addition to the new capabilities outlined above, this level of implementation would require a great deal of aviation and C3 capability in Navy prepositioning ships.
- *Level Three—joint force enabler.* Level Three would add initial joint capabilities and interfaces to the capabilities of the first two levels. For example, it would provide off-load and sustainment support of the Army Regional Flotilla prepositioning ships, use as a temporary Afloat Forward Staging Base for the Army, Special Operations Forces support and employment, C3 for a Joint Task Force commander, port and airfield opening capability, and planning and coordination for joint fires. At Level Three, the sea base would constitute a naval capability.
- *Level Four—full joint integration.* The Level Four capability would permit the tactical employment and sustainment of a wide range of joint forces from the sea base (e.g., Stryker Brigade, Air Assault Brigade, Future Combat System units of action, Army light brigades, Air Force fighter and heavy-lift logistics aircraft squadrons, and integrated joint intelligence, surveillance, and reconnaissance). In this configuration, the sea base would be a fully joint rather than a naval capability.

CONCLUSIONS AND RECOMMENDATION

The committee concludes that Level Four—that is, full joint integration—as described above, is the optimum solution for Sea Basing implementation and recommends that it be adopted as a specific end state. Sea Basing is an inherently joint capability, and it should therefore be planned and designed as such. The committee further concludes that the four levels identified above are distinctly different configurations, and Level One, for example, is not simply a lesser version of Level Three. The Navy needs to define the desired end state and to understand that a spiral development process will not easily progress from Level One to Level Three or Four.

Regardless of the level of Sea Base capability ultimately selected, several programs are common and critical to all implementation levels and should be emphasized. Discussed at length in Chapters 2 through 4, these programs are as follows:

1. High-speed connector development, including the study of a self-deploying, beachable, high-speed connector;[8]
2. Landing craft air cushion X development;
3. Heavy-lift, intratheater airlift[9] (the CH-53X helicopter is the current program of record, but the real requirement is for a heavy-lift (greater than 20 tons) intratheater air transport);[10]
4. Sea State 4 operating capability;
5. Strike-up/strike-down system;[11] and
6. Maritime Prepositioning Force (Future) ship design.

The Department of the Navy widely describes Sea Basing as transformational. The committee agrees with this characterization—if implemented at Level Four as described above, it will enable new alternatives and fundamental change in the way that the Services and the combatant commanders operate. It will have a profound effect on force structure, logistics, training, and infrastructure. The achievement of this capability will necessarily impact a large number of programs and will require significant R&D as well as affecting doctrine, training, and other areas. It clearly requires a management approach that emphasizes systems engineering. The current situation, with concept development, R&D, war gaming, and other activities all apparently proceeding independently (in some cases redundantly and in other cases divergently) will not yield the capabilities envisioned for the Sea Basing levels discussed in the preceding section.

Recommendation: A Joint Sea Base Planning Office—directed by a Navy flag officer or a Marine Corps general officer—should be established. The director of this office should report to an appropriate official in the Office of the Secretary of Defense. The Joint Sea Base Planning Office should be staffed with representatives from the four Services, Special Operations Command, and U.S. Transporta-

---

[8]These applications are of critical importance to the U.S. Army as well as for Sea Basing, and their ability to support the Army Regional Flotilla should be a threshold requirement.

[9]The lengthy assembly/disassembly time of the CH-53E is due to interoperability limitations with current strategic lift platforms and is a significant pacing factor in meeting required deployment time lines. As currently envisioned, the CH-53X program accepts these times and physical limitations as givens. It seems prudent for the CH-53X program to state a requirement for eliminating or for designing-in more efficient or reduced assembly/disassembly times.

[10]Three U.S. military Services (the Army, Navy, and Marine Corps) have similar requirements for the capability to airlift 20 to 23 tons over a 200 to 300 mile radius, originating from existing or planned MPF ships. A joint program office could profitably engage this challenge and might come up with a better solution than a rebuilt CH-53.

[11]Asset-visibility and inventory-control systems need to be developed and fielded for efficient movement, handling, and storage of military items. Also, other design and engineering issues need to be addressed and solved in order to produce a practical, reliable, shipboard cargo-handling system that will achieve the cargo flow volumes and rates that will be required on the sea base.

tion Command. These representatives should be at the O-6 level or above. This office should be responsible for carrying out the necessary studies for the design and operation of the sea base and the airborne and seaborne connectors and for guiding the experimentation with one or more testbeds, as needed. At some appropriate stage of planning, as the studies and experimentation and related Service programs mature, this office could grow into a Joint Program Office for the Sea Base.

The experience of the Services in developing, coordinating, and managing the immensely complicated Joint Strike Fighter (JSF) program is offered as a model for the Joint Sea Base Planning Office. The JSF model is of interest owing to its success in involving all of the Services working through design and budget issues and actually beginning production runs.

2

Long-Range Heavy-Lift Aircraft to Enable Sea Basing

SEA BASING CONCEPT

As discussed in Chapter 1, the Navy and Marine Corps are exploring the issues involved in developing a new capability to deploy from a sea base amphibious forces that would move from the sea base directly to objectives inland. This new tactical concept for conducting amphibious operations is called Ship-to-Objective Maneuver (STOM). The logistics connections required for STOM from a sea base are illustrated in Figure 2.1.

AIRLIFT REQUIREMENTS FOR SEA BASING

If amphibious forces of the future operate according to the vision for Sea Basing, highly mobile Marine units will be widely dispersed well inland when needed. They will be focused on key objectives with high military or political value. They will not be clearing and securing areas through which they pass en route to their objectives.

In some conflicts of relatively low intensity, it may be possible to establish traditional supply lines for moving supplies and equipment and to evacuate the wounded by truck convoy. However, in more typical situations such as in the ongoing war in Iraq, traditional supply lines will be vulnerable to attack from bypassed enemy units, suicide bombers, land mines, and so on. Army and Marine units pushing toward Baghdad, for example, were held up by the need to bring supplies overland through a gauntlet of bypassed Iraqi forces.

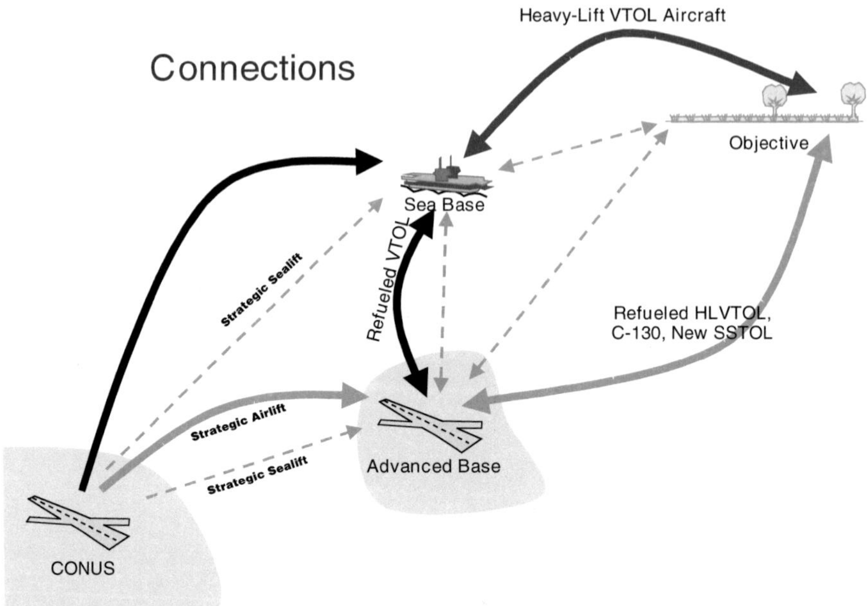

FIGURE 2.1 Maneuver from a sea base will depend on heavy-lift, vertical-takeoff-and-landing (VTOL) airlift. NOTE: A list of acronyms is provided in Appendix C. SOURCE: Constructed from data supplied by Maj Scott Kish, USMC, N703M, "Analyses of Sea Basing Connectors," presentation to the committee, September 9, 2004, Washington, D.C.

The ability to resupply combat forces over long distances without depending on truck convoys is fundamental to the operational concepts of Sea Basing. This means that Marine Corps ground forces and early-entry Army forces will be more dependent on air transport than they ever have been in the past.

The basic requirement for a Marine Expeditionary Brigade (MEB) is the capability to lift one battalion by air and one battalion by surface assets from the sea base within the 8 hour period of darkness. Army requirements for a Brigade Combat Team (BCT) are not available for such a scenario. The minimum threshold value of the payload is 40,000 lb.[1] The objective value is 50,000 lb, based on

[1]The committee assumed an International Organization for Standardization (ISO) standard 20 ft × 8 ft × 8.5 ft container loaded so that the weight of the container plus payload is not over 40,000 lb for the "threshold" value and not over 50,000 lb for the "objective" value. The committee did not discuss the feasibility of smaller supply or vehicular modular units (except for the assembly and disassembly of the CH-53 helicopter).

the weight of the Stryker Interim Combat Vehicle, which is representative of a future light tank. (It is assumed that if heavier armored combat vehicles, such as the M-1 tank, are needed, they would be landed from the sea.) Ongoing studies by the Center for Naval Analyses and the Marine Corps indicate that the distance requirement is an operational radius of 150 to 300 nautical miles (nmi). Greater distances are certainly desirable, perhaps out to 500 nmi, reaching beyond the littoral. The radius and payload capabilities and requirements of the heavy-lift aircraft are illustrated in Figure 2.2 and listed in Table 2.1.

Technical Analysis of Options for Meeting the Airlift Requirements

In many respects, the heavy-lift aircraft will be a design driver for major ships of the sea base, and because of the time that it could take to develop a suitable aircraft, that aircraft could be a pacing item in the overall sea base design. It is therefore worth examining the technical aspects of the design of such an aircraft in some further detail. The following discussion thus examines in more depth the technical details and problems of several options for heavy-lift aircraft.

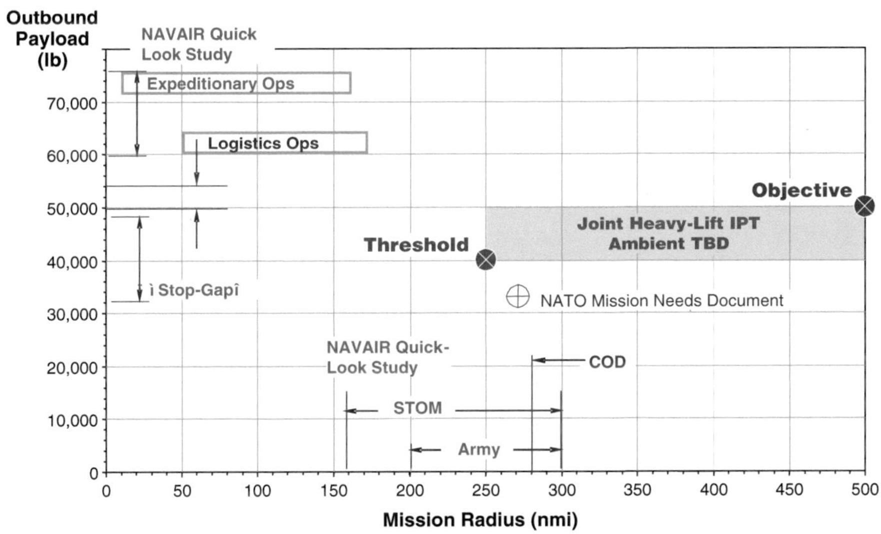

FIGURE 2.2 Heavy-lift, vertical-takeoff-and-landing capabilities: radius and payload requirements to support Sea Basing. NOTES: NAVAIR, Naval Air Systems Command; Ops, Operations; “Stop-Gap,” temporary; IPT, Integrated Product Team; Ambient TBD, ambient temperature and altitude are to be determined; COD, carrier onboard delivery; STOM, Ship-to-Objective Maneuver; lb, weight in pounds; nmi, distance in nautical miles; NATO, North Atlantic Treaty Organization.

TABLE 2.1 Documented Heavy-Lift Requirements for Supporting Sea Basing

	U.S. Marine Corps	U.S. Army	U.S. Navy	Joint Capability
Source for requirements	CH-53X ORD	AMT ICD	DSB Sea Basing and NAVAIR Quick-Look	
Mission	OMFTS/STOM Sustain expeditionary force	Mounted troop transport and sustainment	Access denial Force sustainment VERTREP/VOD/COD	Common heavy lift
Performance				
Payload at radius	27,000 lb at 110 nmi (T) 30,000 lb at 110 nmi (O)	"Mounted" troops FCS ECC "up to 1,000 km"	>20 ton at 300 nmi	24 ton at 270 nmi (FCS at ECC)
Hover time	TBD	TBD	TBD	20 min
Self-deployment distance	TBD	2,100 nmi	2,100 nmi with 1 refuel	2,100 nmi
Takeoff condition	Sea level/103°F	4,000 ft/95°F	Sea level/103°F	4,000 ft/95°F
Midpoint condition	3,000 ft/91.5°F		3,000 ft/91.5°F	
Cruise speed, knots	150 (T), 170 (O)	TBD	180 (A), 240 (G)	>180 kt
Shipboard	LHA/LHD/LHAR compatible 100% (T)/90% (O) footprint	Capable	Compatible Sea State 4	LHD/LHA (R) Compatible Sea State 4
Air transportable	Yes, C-5/C-17	No	TBD	TBD
Internal payload				
Bulk	463L with 10,000 lb each	FCS in ECC	ISO container 30 ft × 12 ft × 10 ft (A) 40 ft × 14 ft × 12 ft (G)	FCS in ECC
Troops	30 crash-rated seats (T) 50 with centerline seats (O)	Yes	TBD	TBD

Survivability	ASE STOM in 1 period of darkness	State-of-the-art ASE	TBD	Active/passive and situational awareness
Operating and support	75% CH-53E O&S costs 100% (T)/90% (O) footprint	TBD	80% availability	Integrated ILS Cost target TBD
Fleet size (est.)	154 (CH-53X ORD)	200 to 512	83	TBD
Schedule	2015 IOC/2021 FOC	FCS increment (2012)	~2015 1st MPF(F) squadron (2017)	FCS increment I (2012)
Other comments	Life: 10,000 to 12,000 KFH (T/O) 10% decrease in logistic footprint		TEU: maximum 26.45 tons MILVAN: maximum 22.4 tons	

NOTES: Threshold (T); Objective (O); Acceptable (A); Goal (G); ASE, aircraft survivability equipment; 463L, military shipping containers are frequently stacked on 463L pallets for storage or transport; ORD, operational requirements document; AMT, Army mounted transport; ICD, initial capability document; VERTREP/VOD, vertical replenishment/vertical ordnance delivery; COD, carrier onboard delivery; ECC, efficient combined container; FCS, Future Combat System; TBD, to be determined; LHA, amphibious assault ship, general purpose; LHD, amphibious assault ship, multipurpose; R, replacement (for LHA); O&S operating and support; ILS, instrument landing system; FH, flight hours; KFH, kilo flight hours; MILVAN, military van.

TABLE 2.2 Performance Characteristics of Current Aircraft That Could Support Sea Basing

Aircraft	Maximum Takeoff Weight (lb)	Payload (lb)	Range (nmi)	Speed (knots)	Internal Height (ft)
CH-47SD Chinook	54,000	28,500	652	140	6.5
V-22 Osprey	47,500	15,000	515	275	6.5
CH-53E Super Stallion	73,500	32,000	110	150	6.5
Mi-26 Halo	123,000	44,000	500	159	~9

SOURCES: Available online at <http://www.boeing.com/>, <http://www.sikorsky.com/>, and <http://www.fas.org/man/dod-101/sys/ac/row/mi-26.htm>, accessed March 9, 2005.

The payload and range performance of current aircraft is summarized in Table 2.2. "Range" is the total distance that can be flown on one tank of fuel, whereas "radius" is the distance that can be flown out and back on one tank of fuel. The radius is generally half the range, unless the aircraft must land and take off again at midmission, in which case the radius can be significantly less than half the range.

With its maximum payload of two Humvees, the operational radius of a CH-53E helicopter is about 100 miles. At a speed of about 110 knots, the CH-53 could transport the assault echelons of an MEB a distance of about 110 nmi in one 8 hour period of darkness. The operational radius of an MV-22, with its maximum external load equivalent to one Humvee, is also about 100 miles.[2]

However, these aircraft are not capable of transporting a light assault vehicle (LAV), each of which weighs about 30,000 lb, over that distance (the Stryker combat vehicle will weigh even more). In addition, an operational radius of 100 miles is not great enough to support the needs of STOM. The maximum payload of the CH-53E and MV-22 decreases with distance, especially with external loads. These two aircraft will not be able to meet all of the needs of Marine Expeditionary Forces over the distances envisioned for Operational Maneuver From the Sea (OMFTS).

Both the Navy and the Marine Corps have considered a notional helicopter designated as the CH-53X with improved performance. However, there are some recognized limitations in the performance that ultimately may be achievable. These limitations are related to the diameter, loading limitations, and tip speeds of the rotor blades. Because of these limitations, the CH-53X has been discussed but is not included in acquisition programs of record or supported by Navy science and technology funding. The Technology Readiness Levels (TRLs) of a notional CH-53X have not been established, but the members of the committee

[2]There are also altitude limitations.

estimate the TRL of the CH-53X concept to be at about TRL 4 (TRL 4 is defined as component and/or breadboard validation in a laboratory environment). The committee estimates that approximately 3 years of research and development (R&D) would be required before design could begin. Optimistically, an initial operational capability (IOC) might be achieved 7 years subsequent to the initiation of procurement. If the CH-53X procurement were to be initiated in Fiscal Year (FY) 2006, an IOC might be achieved by 2015.

The anticipated requirement to carry more than 44,000 lb as far as 300 nmi is similar to the current capabilities of the Russian Mi-26 helicopter, so that the capability to meet the range and payload requirements of STOM has been demonstrated. The existence of the Mi-26 suggests that the technology for such an aircraft is already beyond TRL 6. However, it is not clear that such an aircraft could support the need to deploy the assault echelons of an MEB over 300 nmi in 8 hours, owing to its time en route. Each heavy-lift rotorcraft would be able to make just one round-trip, so that the number of aircraft required would be prohibitve. However, this mission has not yet been analyzed and should be studied.

The committee agrees with the assessment of the Office of the Secretary of Defense[3] that a more modern version of the XCH-62 Tandem Rotor aircraft is the best alternative for a new heavy-lift replacement rotorcraft, because this reduces technical risk compared with that involved in developing a single-rotor aircraft. However, the requirement to transport supplies and heavy equipment over long distances also includes very real operational and survivability considerations. Helicopters are inherently complicated and relatively fragile, requiring up to 50 maintenance hours per flight hour. At altitudes below 8,000 ft, helicopter operations are hampered by weather, dust, and terrain. In addition, ground threats are being keyed toward antihelicopter operations. Enemy interdiction of air routes from the ground will make it difficult to rely on helicopters for support.

For example, the assault on the Medina Division of the Republican Guard near Karbala, Iraq, on March 24, 2003, by AH-64D Apache Longbows revealed the vulnerability of even armored helicopter gunships to small-arms fire. Alerted to the Apaches' approach by lookouts along their route, Iraqi civilians armed only with AK-47s and soldiers with unguided artillery fired barrage-style in a crude but effective ambush. After the failed raid, 27 of the 34 Apaches on the mission were no longer flightworthy. Rotorcraft will have limitations in meeting the air transport requirements envisioned for OMFTS.

---

[3]Discussion between the committee and Michael J. Walsh, Staff Specialist, Office of the Director for Defense Systems, Office of the Under Secretary of Defense for Acquisition, Technology, and Logistics/Land Warfare and Munitions, OUSD/AT&L/LW&M, at the committee workshop, September 8, 2004. Also, see <http://www.flightdailynews.com/farnborough2004/07_20/helicopters/airlift.shtm>. Accessed October 1, 2004.

The development of compound rotorcraft (which utilize wings to develop a significant fraction of their lift in flight), such as the experimental AH-56 Cheyenne, or Gyrocopters and Gyrodynes, would be a step in the right direction. They may be a somewhat faster than helicopters, as demonstrated by the Cheyenne compound helicopter (although similar speeds have not actually been achieved by any Gyrocopter or Gyrodyne). However, these speeds represent only an incremental improvement over those of helicopters. Compound rotorcraft cannot deploy the assault echelons of an MEB over the distances required (150 to 300 nmi) for Sea Basing in an 8 hour period.

In addition, conversion of the Gyrodyne from rotorborne flight to wingborne flight is dangerous because of the need to cut power to the cruise propulsors in order to power up the rotor. During this conversion, the aircraft is falling along a 1 in 4 glide path. Due to the weight and design of such an aircraft, autorotation will not be effective in retarding the fall in the event of failure to start the rotor. It is important to recall that no tipjet-powered helicopters were ever put in service, despite several development programs.

The Air Force Future Air Mobility Command (AMC) is directed toward the development and eventual acquisition of a super-short-takeoff-and-landing (SSTOL) transport aircraft, capable of operating with a balanced field length of 2,000 ft. Even such super-short-takeoff-and-landing distances are too long for the envisioned sea base. The sea base would have to be longer than this to allow for dispersion on landing and braking under the worst conditions—a wet deck with the ship entering the trough of a wave. The wingspan of the SSTOL transport aircraft would be comparable to that of the C-130 (probably more than 130 ft), which would require a flight deck without an island and would prohibit parking another SSTOL transport on the flight deck. Each such aircraft would require that the entire deck be cleared for it to take off and land, limiting the sea base to cycling one aircraft at a time.

There will be other problems involved in the operation of SSTOL aircraft from the sea base. While these aircraft are transferring cargo from the ship to the forces ashore, the cargo on the sea base must be replenished by sealift arriving from an advanced base or from the continental United States (CONUS), as shown in Figure 2.1. The transfer of cargo to the sea base will require the sea base ship to be heading in a direction that protects the transfer from adverse wind and wave effects. On the other hand, launch and recovery of the aircraft will require the sea base ship to be headed into the wind at significant speed. The intricate choreography of heading and speed changes necessary to transfer cargo from the resupply ships, through the sea base, to SSTOL delivery aircraft will make it difficult to support the requirements of STOM in a timely manner. A seaplane is a possibility for resupplying the sea base from CONUS or the intermediate base, but a seaplane would not be suitable for landing at the advanced base.

An extreme STOL aircraft designed to take off and land on the sea base in 500 to 1,000 ft would also be capable of taking off vertically with virtually the same payload as that of the SSTOL. The thrust required to accelerate an aircraft that large to flying speed in those short distances would be sufficient to enable the aircraft to make a vertical takeoff. Similarly, the slow liftoff and approach speeds of such an extreme STOL aircraft (around 75 knots) would require the use of the same reaction control jets that are required for hover. Even with the reduced fuel at midmission, taking off or landing at a remote site in 500 to 1,000 ft and clearing a 50 ft obstacle would probably require the same thrust as taking off vertically from the sea base.

For all of the reasons cited above, the committee believes that the requirements of Sea Basing could lead to the need for a fixed-wing aircraft utilizing some form of powered lift for performance off the sea base ship's flight deck—certainly for SSTOL or STOVL performance, and possibly leading to full vertical-takeoff-and-landing (VTOL) performance, depending on how the aircraft design emerges from detailed design studies.

HEAVY-LIFT TRANSPORT AIRCRAFT ALTERNATIVES

The committee emphasizes the need to pursue advanced technology to develop a ship-capable, fixed-wing, VTOL transport aircraft to meet the requirements of Sea Basing. Efficiency during the 10 to 15 seconds of hover time required for VTOL is not the most important requirement. Cruise efficiency and speed in loading and off-loading cargo are more important. In order to provide for the resupplying and reinforcing of highly mobile Marine Corps, Army, and Special Operations Force (SOF) assault forces operating far inland, the committee recommends the development of the technologies for a ship-capable, fixed-wing, VTOL transport aircraft. As described below, some possibilities include a stowed-rotor aircraft, an aircraft with lift fans in the wings, or an aircraft with thrust-augmenting ejectors in the wings.

Stowed-Rotor Aircraft

A stowed-rotor aircraft would be similar to the stowed-rotor concepts demonstrated in the past, but would carry the concept to the next step by stopping and stowing the rotor in order to achieve significant improvements in speed. The rotor would be slowed, then stopped and stowed in a compartment on top of the aircraft's fuselage, similar to the payload bay of the space shuttle. The folding mechanism would be similar to that developed for the V-22. Stopping and folding a rotor in this way was demonstrated in the National Aeronautics and Space Administration's (NASA's) Ames full-scale wind tunnels more than 30 years ago. The critical technologies are as follows:

- The heavy-lift rotor and transmission system, and
- Integration of the folding mechanism.

Lift-Fan Aircraft

Vertical-takeoff-and-landing operation requires thrust-to-weight ratios greater than that required for cruise. Significant increases in the static thrust of turbofan engines can be obtained by increasing the bypass ratio of the cruise engine for vertical takeoff and landing. The effective bypass ratio can be increased by using the energy in the cruise engine exhaust jet to power a lift fan installed in the wing of the aircraft. The system needs to have the lift fans in the wing, as the fuselage will be taken up with cargo. Similar aircraft, such as the Ryan XV-5, have been capable of taking off conventionally if the lift fans are damaged. Integrating the lift fans with the wing to produce a lift-and-propulsion system would provide a VTOL aircraft that also has good STOL capabilities and increased range and payload performance if they can utilize even a short runway.

The wing lift fans can be driven either by shaft power (as in the X-35B) or by hot gas tip drive (as in the XV-5A). The fans in both concepts would be large in diameter with low fan pressure ratios (FPRs) of 1.08 to 1.20. Previous studies have indicated thrust augmentation (fan lift/shaft horsepower (SHP) or fan lift/thrust) in the range of 2.2 to 2.8 lb per SHP for the shaft-driven fan-in-wing and 2.0 to 2.8 lb per lb for the gas-driven fan-in-wing. The fans would be located at the longitudinal center of gravity with pitch control from fore and aft jets.

Both concepts would have at least two independent engines (one for each side), with cross shafting and cross ducting for one-engine-out capability. Good design practice would have more than two engines to lessen the impact of one engine out and to provide better thrust matching between VTOL and subsonic cruise or loiter. During cruise or loiter, the engines could be powered back or even shut down to match the power or thrust to the cruise or loiter drag.

Using augmentation in the range of 2.2 to 2.8 lb per SHP for the shaft-driven concept gives a power requirement of 25,000 to 32,000 SHP per side. The critical technologies are as follows:

- A lightweight gearbox and clutch to absorb ~25,000 to 32,000 SHP (the current limit is the gearbox for the JSF-35B at 15,000 SHP);
- Turboshaft engines rated at ~10,000 SHP (current limits are the Rolls Royce T 406 and Tyne engines at about 6,200 SHP); and
- Louvers, covers, and structure for the large-diameter wing fan.

Using augmentation in the range of 2.0 to 2.8 lb per lb for the gas-driven concept gives a static thrust requirement of 25,000 to 35,000 lb. The critical technologies are these:

- Louvers, covers, and structure for the large-diameter wing fan;
- In-flight inlet closure;
- Tip-driven turbine seals; and
- Plumbing the equivalent of more than 35,000 lb of thrust in hot, high-pressure gas to the wing-mounted fans.

Ejector Wing Aircraft

Significant increases in the static thrust of turbofan engines can also be obtained by diverting the engine exhaust jet through an ejector, which is a pneumatic device that uses entrainment by the engine exhaust jet to pump a larger mass of air drawn from the atmosphere. A simple ejector consists of a nozzle that directs a jet through a duct. The thrust of the engine is increased by the suction forces that the entrained flow develops on the inlet of the duct. In effect, the ejector functions like a ducted fan and is a mechanically simpler alternative to the lift-fan system. Since ejectors can be used to augment the engine thrust, the additional thrust necessary to give an aircraft VTOL capabilities can be developed from a smaller engine that provides more efficient cruise. Integrating the ejector with the wing to produce a lift-and-propulsion system also provides an aircraft with good STOL performance.

Thrust-augmenting ejectors can also provide the advantage of lower disk loading and a more benign footprint compared with that of a lift-fan system. Mixing of the engine exhaust jet and the entrained air within the ejector duct reduces the velocity, temperature, and noise of the lift jets. The low-temperature and -pressure footprint of this mixed flow would enable an aircraft to operate from ships other than aircraft carriers, and in unprepared, constrained, tactical landing zones ashore. The critical technologies are these:

- Ejector design,
- Enhancement of turbulent mixing, and
- Noise abatement.

Aircraft System of Systems

There may be even more innovative approaches than these described above. The requirements to fly long distances with a heavy payload and to take off and land vertically are almost mutually exclusive. Long-range aircraft must be large in order to carry the necessary fuel, but it is difficult for large aircraft to hover. This is a consequence of the square-cube law, which implies that as the size of an aircraft increases, its weight increases faster than its thrust can be increased. As an aircraft is made larger, its vertical thrust increases only with engine cross-sectional area (L^2), while its weight increases with volume (L^3); therefore, its thrust-to-weight ratio for hovering must decrease. At some size, the propulsion system becomes too heavy for the aircraft to lift. It is for similar reasons that a flea can jump 100 times its own height, while an elephant cannot get all four feet off the ground at the same time.

However, the actual requirement is not to take off and land a large aircraft vertically, but rather to deliver and recover a 40,000 to 50,000 lb payload vertically. Therefore, an alternative approach might be a compound aircraft system consisting of two or more aircraft flying wingtip to wingtip in order to take advantage of the induced drag reductions available from flying in close formation. An aircraft's best lift to drag (L/D) ratio occurs at the speed at which its profile drag equals its induced drag. Because the profile drag of an aircraft is proportional to the ratio of its wetted surface area to its wing area, S_w/S, and its induced drag depends on the aspect ratio of its wing b^2/S, the best L/D ratio depends on its wetted aspect ratio b^2/S_w. The expression is L/D = $(\pi e/4 C_f)^{1/2} b^2/S_w$.

This equation is plotted in Figure 2.3, in which the best L/D ratios of some representative aircraft are spotted on the curve to validate the relationship.

If two aircraft are joined at the wingtips, both the span and surface area are doubled, so that the wetted aspect ratio is also doubled. From Figure 2.3, this can be seen to increase the L/D ratio and range of the paired aircraft by about 60 percent. Joining three aircraft further increases the L/D ratio and range. As long as no aircraft flies in the downwash behind another, all of the aircraft in the formation benefit from formation flying.

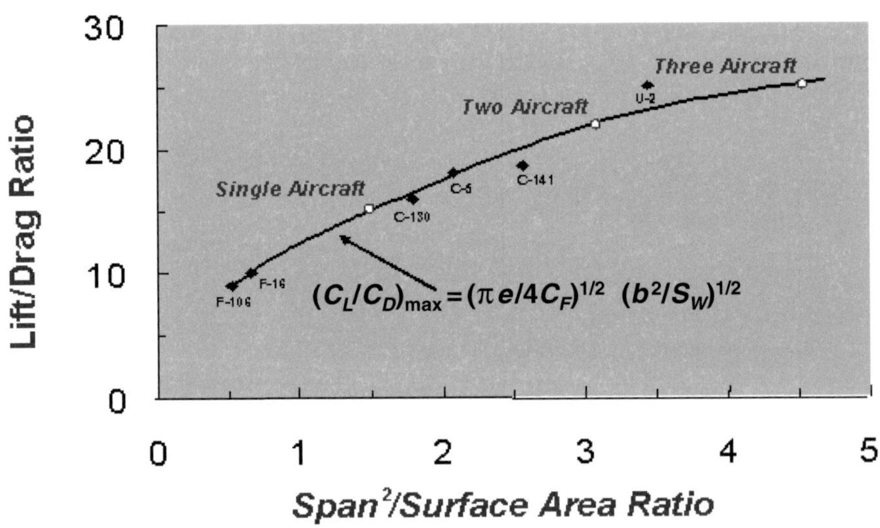

FIGURE 2.3 Formation flight increases lift-to-drag ratio. SOURCE: Paul Bevilaqua, Lockheed Martin Aeronautics Company.

FIGURE 2.4 Project Tom Tom demonstrated drag savings from formation flight. SOURCE: Courtesy of the National Museum of the U.S. Air Force. Available online at <www.wpafb.af.mil/museum/history/postwwii/tomtom.htm>. Accessed June 2005.

In the early 1950s, the U.S. Air Force actually attached and flew a pair of F-84s on the wingtips of a B-36, as part of the Fighter Conveyer program. The exercise was called Project Tom Tom. The test aircraft are shown in Figure 2.4. The hook-up mechanism was similar to the Navy's probe and drogue system. The drag of the system of three aircraft was shown to be the same as that of the B-36 alone, so that the F-84s were ferried with no increase in the fuel required. In terms of the expression in Figure 2.3, the increase in "span^2" that results from attaching the fighters to the bomber was equal to the increase in surface area, so that the L/D ratio did not increase. Air-to-air refueling was selected as a better alternative—it did not seem as dangerous, and fuel was cheap. However, that was the era of bell crank and cables for flight control. With today's fly-by-wire technology, this method of extending aircraft range should be reconsidered.

More recently, the NASA Dryden Flight Research center has validated the performance benefits of formation flight, having measured a 20 percent drag reduction for one F-18 flying behind another, and up to a 60 percent savings for an F-18 flying behind a KC-135. Greater increases in range could be achieved by using one of the aircraft to carry fuel for the others. Since the VTOL delivery aircraft would not have to carry its own fuel, it could be made even smaller. By flying behind and outboard of the tanker aircraft, the VTOL delivery aircraft could realize significant drag reductions, thus increasing range or payload.

Operating from the sea base, two or more smaller VTOL aircraft would take off at the same time and join in formation flight. This could double the range of any of the aircraft flying alone. Greater increases in range could be achieved by using one of the aircraft to carry fuel for the others. Since the VTOL delivery aircraft would not have to carry its own fuel, it could be made even smaller. By flying behind and outboard of the tanker aircraft, the VTOL delivery aircraft could realize significant drag reductions and increase its range. The VTOL delivery aircraft would detach from the tanker at midmission to deliver the payload and reattach for the return flight.

This hitchhiker concept would reduce the technical risk and cost associated with developing heavy-lift VTOL aircraft to support Sea Basing, by increasing the range of smaller aircraft that can more easily make vertical takeoffs and landings. The critical technologies are these:

- The VTOL lift system, and
- The software for automatic formation flight.

DEVELOPMENT STEPS

Concepts for all of the heavy-lift VTOL transport aircraft discussed in the preceding section are at TRL 2. Technology for heavy-lift aircraft would be matured to TRL 6 in four phases of increasing scope and complexity. The first phase would consist of analytical predictions of lift system and airplane performance in hover and transition, together with small-scale, wind tunnel model testing of airplane configurations in hover and transition. This work would bring the concepts to TRL 3.

The second would consist of a test of a full-size propulsion system powered by an aircraft engine in order to provide data at the full-scale lift jet Mach number and Reynolds number. These tests would be conducted on a whirl rig to investigate transition to wingborne flight. This work would bring the concepts to TRL 4.

In the third phase, a large-scale airplane model would be built and suspended from a gantry for hover testing. It would be mounted in the NASA Ames 80 ft × 120 ft wind tunnel to show transition characteristics. This work would bring the concepts to TRL 5.

In the fourth phase, to bring the concepts to TRL 6, it might be possible to replace the wing on an existing transport aircraft, such as the C-130 Hercules or C-27 Spartan, with a new wing. The aircraft would be flown from a hover through transition to wingborne flight in order to demonstrate successful development of the control laws, the ability to generate adequate control power, and the VTOL performance of a heavy-lift transport aircraft.

A very rough cost and schedule estimate for such a program is shown in Figure 2.5. The costs shown are not cumulative; each TRL level has been esti-

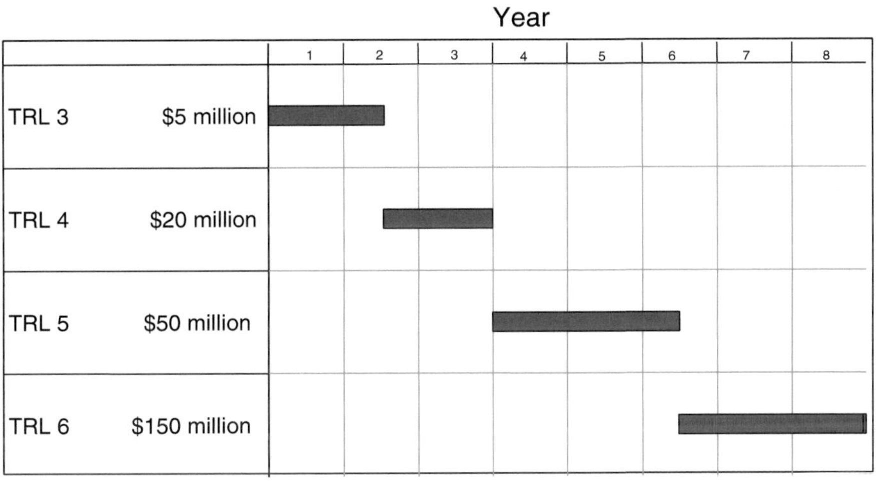

FIGURE 2.5 Very rough schedule for enhanced heavy-lift aircraft development, in years.

mated separately, but the estimate depends on the earlier levels having been accomplished. These estimates reflect the level of effort expended on similar development programs in the past. However, the very rough scope and phasing of these tasks should be considered useful only to begin discussion of the priorities and goals of Sea Basing.

CONCLUSIONS AND RECOMMENDATIONS

The committee concludes the following:

- Given a heavy-lift replacement for the CH-53E helicopter and acquisition of the V-22 tilt-rotor aircraft, the Marine Corps will have the capability to move substantial quantities of materiel by air. However, these two aircraft do not have the range, payload, and speed performance to meet all of the needs of the assault echelons of a Marine Expeditionary Brigade over the 150 to 300 nmi distances and 8 hour deployment times envisioned for Operational Maneuver From the Sea, nor to meet the requirements for an Army brigade-level force of some strength, to be determined, for operation from the joint sea base. Therefore, the committee believes that the requirements of Sea Basing could lead to the need for a fixed-wing aircraft utilizing powered lift for VTOL performance.
- Powered-lift system concepts for such a heavy-lift VTOL transport aircraft are currently at TRL 2. Maturing the technology for each preferred

powered-lift system to TRL 6 could be accomplished in four phases of increasing scope and complexity. This will require up to 8 years and at least $150 million for each concept. The required time could be reduced with a corresponding increase in risk.

Recommendation: The use of advanced technology should be pursued to develop a ship-capable, fixed-wing aircraft having powered lift of some as-yet-undetermined configuration with approximately the payload capacity of a C-130J, and that the aircraft will be able to operate in super-short-takeoff-and-landing (SSTOL) or short-takeoff-and-vertical-landing (STOVL) mode, and possibly in full vertical-takeoff-and-landing (VTOL) mode. Such a transport aircraft should have the capability to carry a standard International Organization for Standardization (ISO) 20 ft container or the Stryker combat vehicle to an operational radius of 150 to 300 nmi at high speed and altitudes in order to meet the requirements of Sea Basing. The development of such an aircraft should be undertaken as a joint Service program in collaboration among the Air Force, Army, Navy, and Marine Corps.

Recommendation: The Joint Heavy-lift Aircraft Exploration of Concepts being coordinated within the Office of the Secretary of Defense should involve the U.S. Transportation Command in the process. This effort should be transferred to the Joint Sea Base Planning Office when this office is created.

3

New Ship Classes for Sea Basing

BACKGROUND

As discussed in Chapter 1, the goal of Sea Basing—one of the three fundamental concepts underlying Sea Power 21—is to project joint power from the sea. Sea Basing is designed to use the sea as a maneuver space, to give the Joint Task Force commander the means to achieve accelerated deployment and employment times, and to enable joint follow-on forces from a mobile platform to operate unencumbered by host-nation requirements. Analysis is ongoing to determine the correct mix of assets for the deployment of these forces to the objective. The analyses to define the ship types and classes that were reviewed for this report were focused on the sea base and the strategic and tactical connectors associated with the major nodes. As discussed in Chapter 2 (see Figure 2.1), the Sea Basing concept has four major nodes: the continental United States (CONUS), the advanced base, the sea base, and the objective or shore. A simplified illustration of the Sea Basing concept is shown in Figure 3.1.

In this chapter, the committee reviews ship classes or types as methods of connecting the nodes. A discussion of joint operations is also included.

CLASSES OF SHIPS

Continental United States to Advanced Base

Strategic air transportation is the primary means of moving troops and some equipment from CONUS to the advanced base during time-phased, force deploy-

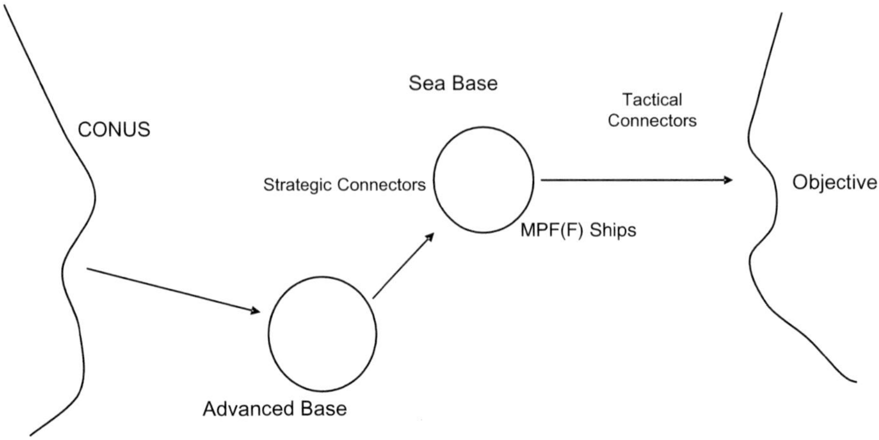

FIGURE 3.1 Simplified illustration of the Sea Basing concept. NOTE: A list of acronyms is provided in Appendix C.

ment. Analysis is also being conducted on a secondary means, high-speed sealift (HSS), as a method of transporting the non-self-deploying rotary wing aircraft. The concern is that the MH-53 helicopter and the subsequent CH-53X require disassembly before and reassembly and testing after transport within the C-17 Globemaster III aircraft. The HSS is being investigated to address this concern, with long-range, high-speed vessels of sufficient payload capacity to carry the MH-53 or the CH-53X from CONUS to the advanced base, or to the sea base directly, without disassembly. However, in the planning stages, care must be taken to assure that the operational requirements for the HSS do not transcend what is physically realistic.

Advanced Base to Sea Base

The advanced base is defined as a sea and aerial port lying within 2,000 nmi of the sea base; through it the troops, equipment, and supplies will flow into the sea base. The method of transport of the troops, equipment, and supplies to the sea base is highly dependent on how far the sea base (or the Maritime Prepositioning Force; MPF) is from the advanced base. Once it is outside of the range for the non-self-deploying helicopters, surface connector technology must carry troops, equipment, and supplies and have the ability to interface with the sea base ships. Ships of the combat logistics force (CLF) may be required for transferring cargo from CONUS or the advanced base to the ships of the sea base/Maritime Prepositioning Force (Future) (MPF(F)) ships. The survivability and transfer requirements of the CLF and MPF(F) ships need to be addressed.

Sea Base Composition

MPF(F) ships are considered to be the main platforms of the sea base, although conceptually other alternatives exist. The Navy's Analysis of Alternatives is complete for the MPF(F), but the Navy, Army, and Marine Corps have not settled on either their own requirements or possible joint requirements for the ship. In any case, to ensure that cargo can be transferred in a Sea State 3 or 4, the MPF(F) ships will have to be larger and will need motion-control technology and self-sustaining cargo handling for both lift on/lift off (LO/LO) and roll on/roll off (RO/RO) cargo. As the MPF(F) becomes the center of gravity for the amphibious forces, issues of survivability will need to be resolved; these will relate both to the sea base protection provided by Sea Shield and to the ship survivability standards incorporated into the MPF(F) itself. Additionally, other issues will need to be investigated—for example, the number and types of surface interface points such as the integrated landing platform (ILP) or the mobile landing platform (MLP); the need to configure the MPF(F) ships with a flight deck to accommodate the heavy-lift aircraft, described in Chapter 2, with power-assisted takeoff and landing or electromagnetic catapult assists for purposes of transferring troops and high-value cargo; or the need to carry landing craft, air cushion (LCAC)- or LCAC-X (experimental LCAC)-type connectors. Lastly, the conflicting capabilities of cargo transfer in high sea states versus the aviation flight capacity needed for the air assault portion of the amphibious operation will need to be resolved in order to permit a clearer definition of the sea base platform requirements.

Maritime Prepositioning Force (Future)

Currently, a total of 36 military cargo ships, manned by civilian crews and operated by the Military Sealift Command, store pieces of military materiel and logistical supplies for various parts of the U.S. armed forces. The Maritime Prepositioning Force, comprising 16 of these ships, carries equipment and supplies for the Marine Corps; the 10 ships carrying Army equipment are called the Combat Prepositioning Force; the remaining 10 ships carry equipment and supplies for the Air Force, the Navy, and the Defense Logistics Agency.

The replacements for the ships that support the Marine Corps are called the MPF(F), and although their design is as yet undecided, the Navy has indicated that they will have the following new capabilities, not possessed by today's MPF ships:

- *At-sea arrival and assembly.* With this capability, joint ground forces could marry up with their equipment and supplies at sea rather than relying on the availability of the friendly ports required by today's MPF; and
- *At-sea selective onload/offload or strike up/strike down.* With these capabilities, MPF(F) ships would provide selective loading/unloading at sea of specific

equipment or cargo without having to off-load other items of unneeded equipment or cargo in order to facilitate force tailoring and sustainment of forward land operations indefinitely from the sea base.

Although specific designs are still undecided, it is recognized that the future MPF would most likely feature at least two nominal classes of ship: MPF(F) vessels and the Maritime Prepositioning Force (Aviation) (MPF(A)), a ship with increased aviation capability. The MPF(F) and MPF(A) vessels would collectively be significantly more capable than today's ships are, bringing a new Sea Basing capability to the joint force. Their costs, depending on the approach taken, could range from $1 billion for a ship designed primarily to commercial survivability standards, to more than $3 billion or $4 billion for what is described as a larger, multimission "distributed" ship built to primarily military survivability standards.

The Army is considering several options for updating its current Combat Prepositioning Force fleet and intratheater shipping. The options include a new Afloat Forward Staging Base (AFSB) capability and an assortment of high-speed intratheater surface connectors that could be built to Army specifications, built to commercial standards, or converted from existing large-displacement commercial cargo vessels. The realization of these capabilities should logically be closely linked to the larger joint sea base effort and could even include common hull forms.

Regarding the MPF(F), the committee concludes the following:

- The presentations given before the committee indicated a lack of uniformity or consensus among the briefers regarding their organizations' understanding of exactly what level of hostilities or spectrum of conflicts the vessels and connectors of the future Sea Basing fleet are meant to contend with. Whether these vessels are intended to operate only in low-threat or Sea Shield-protected environments or are to be capable of conducting independent operations in contested waters will have a significant impact on the standards of survivability, organic self-defense capability requirements, civilian or military manning decisions, and ultimately the cost of building such a capability.
- Deciding whether to build some MPF(F) ships to primarily civilian commercial standards or only to military warship standards is one of the most important decisions to be made about the future Sea Basing program options. This decision will have broad implications for the total costs of concept implementation. Additional considerations include legal issues surrounding questions of civilian or military manning and the question of whether or not there will be common standards across the other vessels being planned to support the Army and Air Force components of Sea Basing.
- Given the likely constraints on the Navy shipbuilding budget and future force structure, any decisions on the MPF(F) and MPF(A) vessels regarding such

matters as increased displacements and multimission-enhanced capabilities for aviation operations, command and control, fires integration, sea-based hospitals, organic connectors, and so on may well require some significant trade-offs with the more traditional military programs. As an alternative, the Navy may request supplemental funds, but such funding would have to be available over a long enough period to complete the acquisition of the sea base. The Navy should also consider having the private shipping industry bid on developing MPF(F) ships for lease or sale.

- Currently the MPF(F), Army AFSB, potential intratheater connectors—such as theater support vessel (TSV), landing craft support (LCS), and the Marine Corps heavy-lift helicopter CH-53(X)—are treated as totally independent programs with separate missions and built to independent standards. Under a joint Sea Basing concept, these programs need to be joint under common joint Sea Basing standards from their conception on so as to be not only interoperable, but also optimized to perform in a complementary manner in the future joint sea base operating environment.
- The characteristics of the MPF(F) cannot be defined, nor can its design be developed until the required ship capacity and cargo throughput are determined, as well as the methods of on- and off-load at the interfaces with the supply connectors and with the shore-bound connectors.

Expeditionary Strike Groups

Expeditionary Strike Groups (ESGs) are a new type of naval tactical formation composed of amphibious warfare ships, surface combatant vessels, and submarines, and supported by land-based P-3 aircraft. Prior to the creation of ESGs, the Navy's amphibious ships were organized into 12 Amphibious Ready Groups (ARGs). The Navy is now converting all of its ARGs to ESGs. The amphibious warfare ships assigned to ARGs and ESGs are built to survivability standards similar to those for other Navy battle force combatants and are manned by Navy crews with Marines as passengers. Such ships include amphibious assault ships, general purpose (LHAs); amphibious assault ships, multipurpose (LHDs); landing ships, dock (LSDs); and amphibious assault transports, dock (LPDs). The current force of 36 amphibious ships has the capability of embarking two Marine Expeditionary Brigades of approximately 13,100 Marines each; however, at any given time 15 to 20 percent of this force is in for maintenance or overhauls. Regarding ESGs, the committee concludes the following:

- Future decisions regarding the capabilities, numbers, and composition of the Expeditionary Strike Groups of 2015 will have a significant impact on the required capabilities, numbers, and cost of the MPF(F) ships. A complete definition of future ESG characteristics is required, along with a joint Service capability (interdependence) analysis, in order to conduct the capability gap analysis to clearly define required MPF(F) capabilities.

- Absent any national commitment to building a larger Navy or increasing the Ship Construction, Navy (SCN) budget above current levels to achieve a more capable fleet, future investment in traditional amphibious ships is likely to be impacted by an increased investment in achieving a new joint Sea Basing capability.

Connectors

After the decisions are made regarding the future ESG composition and the capabilities required of the MPF(F), the next challenge can be addressed: that of providing the appropriate mix of self-deploying and organic surface and air connectors for sustaining the sea base and successfully projecting a significant joint ground force to the objective from the sea base and sustaining this force indefinitely. Although there appear to be some interesting achievable technology alternatives such as the landing craft support, TSV, beachable high-speed connector (B-HSC), and some emerging heavy-lift vertical-takeoff-and-landing (VTOL) technologies, the committee found that the individual Services are mostly planning to pursue improved models of existing or planned models of LCAC, heavy-lift helicopter (CH-53X), and tilt-rotor aircraft (V-22).

Two critical decisions that will have a strong influence on the design of the high-speed connector (HSC) must be made:

- Will the shore-bound HSC be beachable, or will it be an LCAC shuttle, carrying loaded LCACs close to the beach? The latter approach provides amphibious capability so that cargo can be off-loaded ashore above the high-water mark or even further inland. This ability is a major advantage. The decision on this issue will have a major impact on the shore-bound HSC size and hull configuration.
- Should the shore-bound HSCs and the advanced-base-to-sea-base HSCs have a common design or be entirely different classes? A common design would likely have cost advantages if it proved to be possible.

Both of these decisions, along with many others, should be the subjects of trade-off studies (cost, performance, risk) before decisions are made.

Regarding connectors, the committee concludes the following:

- To date, the individual Services have exhibited little progress in pushing the technology envelope to seek new, improved designs of heavy-lift aircraft and high-speed surface connectors. Instead they have focused their work on updated versions of legacy programs. This fact has resulted in a divergence of approaches and little progress in achieving consensus on a joint sea base concept.

- The Services appear to have dismissed any consideration of more revolutionary technology for unmanned connectors and have not conducted any significant analysis to support decisions regarding a best mix of self-deploying or MPF(F)-carried or organic connectors and the consequences of either choice on closely related programs. The Services need to consider unmanned connectors and to provide a transition path to unmanned connectors if they are to be recommended.

Sea Base to Objective

A number of factors will have an effect upon the sea-base-to-objective connector class. These include—

- The type and number of surface interface points for the MPF(F) ship,
- The distance of the sea base from the shore, and
- The need to beach the connector to have it go farther over the beach.

Each of these factors will affect whether the emphasis is placed on ships such as the beachable high-speed connector or an experimental air-cushioned vehicle, the LCAC-X, or whether the emphasis remains on legacy connectors. A beachable transit connector imposes some challenging constraints on the design. However, it appears that certain connectors need to be beachable. Given that the purpose of the sea base is to be just that, the design of the open-ocean, high-speed connectors is greatly simplified if the beaching capability is provided between the sea base and the shore (LCACs, landing craft, utility (LCUs), and so on).

Joint Operations

The committee was briefed on the Afloat Forward Staging Base concept by Maersk, Inc.; the AFSB could be used by the Army as an adjunct to the Navy's Sea Basing concept. As discussed in Chapter 1, the Sea Basing concept has limitations and at its maximum capability, Level Four—Full Joint Integration, it would be used as the naval contribution to joint operations. Under Level Three—Joint Force Enabler, joint operations under a Joint Task Force commander conducted from the sea base would likely require the Army to be supported by an AFSB concept.

Surface Connector Node Map

Figure 3.2 maps out the Marine Corps vision for the Sea Basing connector classes that need to be investigated in order to take full advantage of the Sea Basing concept. The figure indicates the main nodes of the path from CONUS, advanced base, Maritime Prepositioning Squadron (Future) (MPSRON(F)) as the

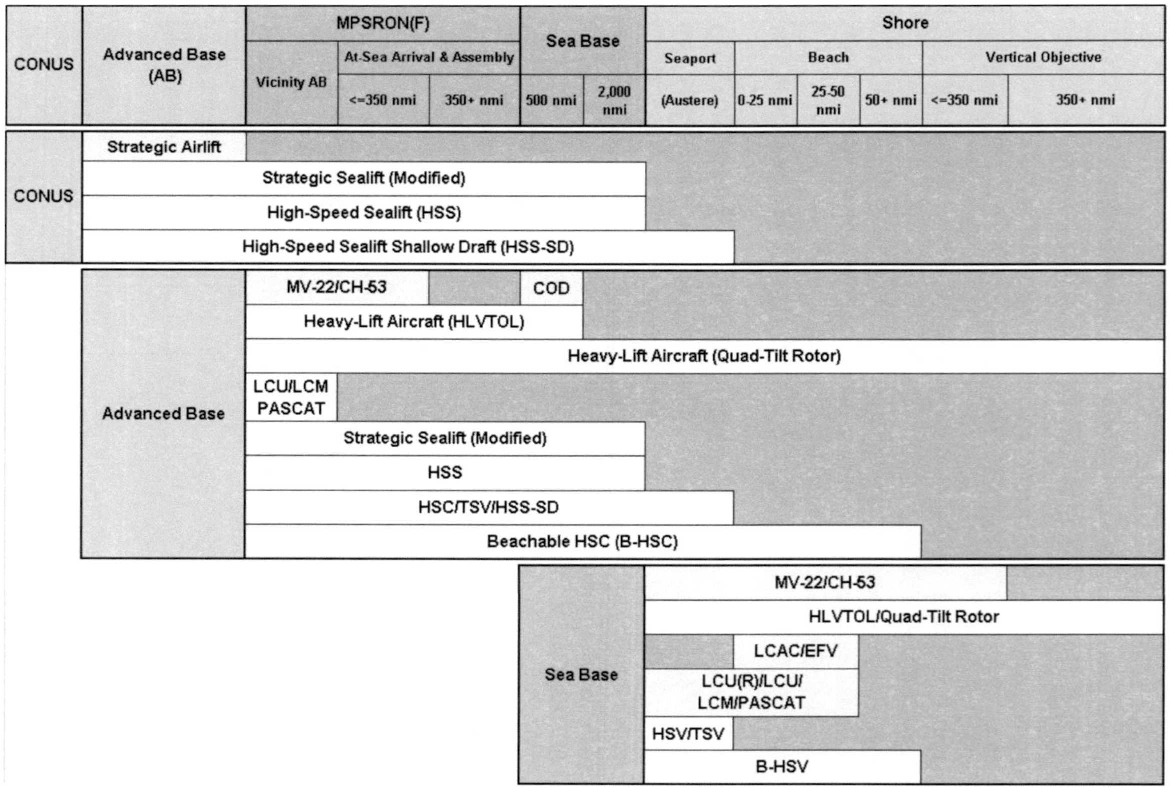

FIGURE 3.2 Marine Corps vision for the Sea Basing connector classes. NOTE: A list of acronyms is provided in Appendix C. SOURCE: Maj Scott Kish, USMC, N703M, "Analyses of Sea Basing Connectors," presentation to the committee, September 9, 2004, Washington, D.C.

connector moves toward the sea base, the sea base itself, and finally, the shore. The figure is further broken down as to the distance that the connector must traverse. For the MPSRON(F), this would include the vicinity of the advanced base (where short-range connectors such as the landing craft, utility, or the British partial air cushion supported catamaran (PASCAT) could be used. An example of using the figure would be that the means of connecting the advanced base to the MPSRON(F) would be MV-22 tilt rotor or the CH-53 helicopter if the MPSRON(F) was within the vicinity of the advanced base or within 350 nmi if conducting at-sea arrival and assembly.

CONCLUSIONS AND RECOMMENDATIONS

The committee concludes the following:

- *Requirements.* There appears to be a desire to push the limit on requirements beyond what engineering can accomplish, given the real limits of physics. For example, seaborne connector speeds of 40 knots or more may be achievable in low sea states, but at great cost and risk. Similar issues exist for cargo transfer at sea and the beaching capability of larger seaborne connectors.
- *Sea base air capability.* Until recently, little consideration appears to have been given to providing capability for the launch and recovery of an aircraft that can carry the load capacity of a C-130J within the sea base; initiatives are currently underway in the Office of the Secretary of Defense to address the heavy-lift aircraft alternatives.[1] The MPF(F) ships have missions independent of a sea base that make their configuration incompatible with incorporating takeoff and landing capability for such aircraft.
- *Coordination.* Despite great similarities in the objectives of various Services, there appears to be a lack of coordination on related programs even within the Department of the Navy. Because of the significant warfighting implications of a sea base, coordination among all Services should be enhanced.
- *Ship motions and cargo transfer.* The physical realities of wave-induced ship motions on all types of ships indicate that only very large ships may even hope to yield successful platforms for effective cargo transfer at sea in Sea State 3 or 4. The committee believes that unless a large testbed (possibly an LMSR (large, medium-speed roll on/roll off)-size ship) is made available to test engineering designs for improved high-sea-state cargo-handling concepts, new cargo-handling capabilities are unlikely to be fielded by the Navy within the next 10 to 20 years.
- *Development.* MPF(F) ships must be designed in concert with the vertical-take-off-and-landing/short-take-off-and-landing/super-short-takeoff-and-vertical-

[1]These initiatives include the Joint Vertical Aircraft Task Force and the Army-led Team on Heavy-Lift Aircraft for Seabasing.

landing technology capability for the MPF(F)'s initial operational capability date.

- It would be premature to push ahead with the MPF(F) design and procurement until certain critical decisions are made. These decisions are dependent upon a total systems analysis from which the required MPF(F) cargo capacity and cargo-transfer throughput can be derived.
- The MPF(F) ship design will be driven by the cargo-transfer techniques adopted for it and by the number of transfer stations required to meet the required cargo-transfer rate.
- The high-speed surface connector designs will also be driven by the required cargo flow rates and the cargo-transfer techniques adopted for the MPF(F).
- There are strong interdependencies between the HSC and the MPF(F) designs; these critical system elements are linked by the required cargo flow rates and the selected cargo-transfer techniques. These ship types must not be developed independently.
- Key issues that will drive overall system cost and performance as well as the designs of the HSC and the MPF(F) are the development and approval of the concept of operations for the sea-base-to-objective connectors as they fit within the context of the larger, Sea Basing concept of operations. A systems analysis of the sea base, as a node, to the objective, as another node, should define the designs.

The committee recommends the following:

Recommendation: A comprehensive systems analysis of Sea Basing ships and connectors needs to be undertaken on a macro level to validate what the requirements (range, speed, and capacity for cargo and personnel) should be. This is especially true of the speed requirements for the various connectors between the Sea Basing nodes. Consideration should be given to using a separate aircraft capable ship for operations of a fixed-wing, vertical-take-off-and-landing, super-short-takeoff-and-landing, or short-take-off-and-vertical-landing aircraft of C-130J size. This capacity should be considered as a separate part of the systems analysis. If this air capability is needed, the Air Force must be part of the development team because of its experience in acquiring and fielding aircraft of this size and capability.

Recommendation: Future developments and analyses of Sea Basing ships and connectors should be conducted under the leadership of a flag-level joint analysis team, which may eventually become a joint program office. Separate stovepipe Service programs should be rolled, as applicable, into this joint program. Owing to its proven capability in ship procurement, the Navy should have the lead in the acquisition of the final ship and connector designs for Sea Basing, which ultimately should be common for all Services.

Recommendation: The Department of the Navy should identify one large vessel to be used as a testbed for resolving the known problems, including those related to connectors and internal cargo handling, involved in at-sea cargo transfer at Sea States 3 and 4, or two such vessels if required for an integral flight deck in order to explore issues associated with potential future heavy-lift aircraft. The Department of the Navy should pursue private industry proposals for the acquisition of larger commercial vessels for such testbeds.

The Department of the Navy could use an existing large ship—a large, medium-speed roll on/roll off or an existing Maritime Prepositioning Force-type reserve ship in a reduced operating status, such as an SL-7/T-AKR Fast Sealift Ship—as a test platform to test advanced engineering development models aimed at the effective sea transfer of cargo in Sea State 3 or 4. It may be necessary to activate a reserve carrier or an equivalent ship for experimentation with heavy-lift aircraft at sea.

4

Open-Ocean Cargo Transfer to and from a Sea Base

BACKGROUND

As indicated in Chapter 1 of this report, the Sea Basing concept has not been stated with sufficient definition to allow formal publication of the required capabilities for cargo handling and at-sea transfer of loads that must be incorporated into the selected platform or platforms that will constitute the sea base. On the basis of worldwide data provided by the Oceanographer of the Navy on the frequency of occurrence of various sea-state conditions, the Defense Science Board (DSB) report on Sea Basing recommended that an open-ocean cargo-transfer capability (to and from a sea base), in conditions up to 3 and 4 Sea States should be achieved.[1] This goal was considered desirable so that the percentage of time during which operations might be limited by unfavorable environmental conditions will be relatively small. The committee believes the DSB sea-state performance goal to be reasonable. The DSB report implies that the types of cargo that a sea base should be able to receive or disperse in such sea states would include 20-ft-equivalent unit (TEU) shipping containers, outsized/heavy (greater than 40,000 lb) equipment, pallets, and liquids (fuels and/or water).

To some extent, current commercial and military capabilities permit such transfers in nonport environments during times of relatively benign sea states. As

[1]Defense Science Board. 2003. *Defense Science Board Task Force on Sea Basing*, Office of the Under Secretary of Defense for Acquisition, Technology, and Logistics, Washington, D.C., August. The committee notes that the probability of sea-state conditions and their duration will be an important factor that will influence the sea base design and operational effectiveness.

an example, the ship-to-ship transfer of TEUs in nonbenign conditions is difficult, because in commercial practice the TEUs must be stored within guide rails that leave little clearance even for minor deviations from the vertical caused by pendulous motion resulting from the pitch or roll of the vessels executing the transfer. The problems of cargo transfer at sea would be immensely simplified if all of the Services adopted common joint pallet and container sizes. While 20 or 40 ft International Organization for Standardization (ISO) containers will normally be employed for supplying or resupplying the sea base, containers that are smaller than the commercial ISO containers are generally needed by the troops fighting in the objective area. The committee believes that the Joint Modular Inter-Modal Container (JMIC) program is an important initiative that, if successful, will alleviate many of the problems currently encountered in this regard. A number of JMIC containers can be arranged and locked together into 20 or 40 ft ISO-container-size blocks for transport in commercial containerships.

The offshore construction industry has lifted and transferred loads of 5,000 tons or more at sea, using built-for-the-purpose equipment. However, such activities are infrequent and have generally taken place during periods of benign sea states. Offshore resupply boats (mud boats) operate routinely in Sea State 3 or even higher sea-state conditions. However, to the best of the committee's knowledge and experience, mud boats are not used for the transfer of loads comparable in weight to militarily significant loads (e.g., 70 ton tanks) during periods of high sea state. Also, during mud boat operations, one of the platforms involved is fixed in space, making the transfer easier than it is when both ships involved are subject to random uncorrelated motions.

Various studies of the Sea Basing concept have considered solutions to the open-ocean cargo-transfer problem that include some combination of the use of the following:

- Vertical-lift or short-takeoff-and-landing (STOL) aircraft,
- Stabilized cranes for skin-to-skin transfers,
- Mobile landing platforms (MLPs) (float-on/float-off heavy-lift ships),
- Wire line and hose transfers,
- Integrated landing platforms (ILPs),
- Stern elevators,
- Roll-on/roll-off ramps, and
- Transfers within a well deck into air cushion vehicles (landing crafts, air cushion (LCAC) or LCAC-experimental (X)).

Each of the techniques in the foregoing list is proven technology in commercial and/or military cargo-transfer operations. All of them have significant performance limitations, and the limitations in their performance are correlated with the attributes of the ships and connector platforms involved as well as with the attributes of the sea states.

The importance of intraship cargo handling on the sea base ships should be emphasized. Cargo loads brought to the sea base must be broken down and reconstituted into tailored packages for transfer to the troops in the objective area. Marine shipboard cargo-handling systems, including the processes of strike-down, stowage, strike-up, breakout, repackaging, and so on, should be developed with the ability to sustain the required cargo flow rates. Full-scale at-sea testing will be required to develop the required systems.

Ideally, the selected platform design for a sea base should incorporate inherent capabilities to accomplish the following:

- Provide for the rapid transfer of large tonnages of cargo both from arbitrarily configured container ships and tankers and from those designed for the purposes of naval vessels (e.g., AO, AKE, AOE,[2] and so on);
- Provide a high strike-down rate for received cargo;
- Provide rapid and effective transfer of equipment to a variety of sea connector designs (e.g., LCAC, LCAC-X, beachable high-speed connector (B-HSC), heavy-lift LCAC (HLLCAC), proposed amphibious helicopter assault vessel, high-speed vessel (HSV)/theater support vessel (TSV), and so on);
- Allow the selective retrieval of any specific item of stored equipment;
- Keep position in conditions up to Sea State 4 and ballast down sufficiently to reduce roll to a small fraction of a degree;
- Transport and provide operational support for non-self-deployable heavy-lift vertical-takeoff-and-landing (VTOL) aircraft (including Army variants); and
- Accommodate the landing and takeoff of STOL aircraft with payloads greater than 20,000 lb.

Any assessment of the technology that will be necessary to achieve the desired cargo-transfer and -handling performance goals for a sea base is inherently dependent on the assumptions that are made concerning the design and attributes of the sea base, the sea-base-to-shore connectors, and the logistic resupply ships and aircraft that will service the sea base. Critical design factors will include such things as the following:

- Hull size and shapes of the sea base, the sea-base-to-shore connector vessels, and the logistic resupply vessels that service the sea base;
- The availability and effectiveness of roll-mitigation systems;
- An ability to operate in deeply ballasted conditions that minimize roll, pitch, and heave motions;

[2]AO, Auxiliary, oiler; AKE, Auxiliary, cargo and ammunition ship; AOE, Auxiliary, oiler and supply ship (fast combat support ship).

- Deck designs, including deck strength, the location of obstructions to STOL operations, the availability of electromagnetic catapults,[3] and so on;
- Elevator-lift capacity and internal cargo-handling and -management systems that will permit the rapid retrieval of any stored item of equipment;
- Dynamic position-keeping capabilities; and
- The cargo-transfer methods that are adopted (dry wells, floating platforms, outboard elevators, and so on).

The committee notes that the designs currently proposed for a sea base or for the Maritime Prepositioning Force (Future) (MPF(F)) do not incorporate all of these capabilities or attributes. The interdependencies between the core element of the sea base, the MPF(F), the supply connectors, and the shore-bound connectors must be emphasized. The interfaces between these two categories of connectors and the sea base itself (i.e., the cargo-transfer techniques to be employed) will have a major influence on the MPF(F) configurations as well as on the designs of the several connectors. The MPF(F) design should not be frozen until the required cargo flow rates have been defined and the cargo-transfer methods to be employed have been selected and demonstrated in full-scale, at-sea tests in demanding environments. As an example, if either a stern elevator or a stern ramp is required for the MPF(F), its inclusion will have a major effect on the ship's overall general arrangement.

CURRENT STATUS OF THE SCIENCE AND TECHNOLOGY BASE AND TECHNOLOGY GAPS

Although limited to lower sea states (below Sea State 3), most of the cargo-transfer capabilities desired for a sea base exist in commercial and/or naval practice. The committee believes that the extension of current open-ocean (nonport) cargo-transfer capabilities to use in higher sea states may be viewed as a number of difficult (but tractable) engineering problems that can be resolved with the application of extant technology and engineering techniques. Few of the current constraints require an extensive science and technology (S&T) investment for resolution. Almost the entire current Navy S&T or research and development (R&D) effort that is oriented toward the achievement of at-sea cargo-transfer capability in higher sea states is vectored toward inclusion in the MPF(F) design. Unfortunately, design efforts for the MPF(F) do not assume that such Navy S&T or R&D efforts will be successful prior to the preconstruction design freeze for the MPF(F). Vaguely stated plans exist for retrofitting or for inserting improved cargo-handling capabilities into current or future MPF(F) hulls.

---

[3]The power requirements for catapults will be a significant factor in the design of the ship's power plant.

The committee is concerned that no integrated development plan seems to exist that supports an R&D effort specifically designed to provide high-sea-state cargo-transfer capabilities for the MPF(F). The following subsections present the committee's assessment of current cargo-handling capabilities and its prognosis for the eventual achievement of significantly improved capabilities in either an MPF(F) or a sea base.

Vertical Lift

Aviation is a key element in the array of cargo-transfer mechanisms from the sea base to shore, whether to a beachhead or to forces maneuvering against an inland objective. A detailed discussion of the technical issues in developing heavy-lift aircraft to transfer cargos on the order of 20 to 23 tons from the sea base to shore is presented in Chapter 2. As indicated in that chapter, the technical requirements of range and payload to be moved even from a very large ship will require an aircraft having some kind of power-assisted takeoff and the capability for vertical landing.

This chapter will explore the technical problems of transferring cargo among ships of the sea base, and from the sea base to connector ships that will take the cargo to shore.

Stabilized Cranes for Skin-to-Skin Transfers

Stabilized cranes for skin-to-skin cargo transfers exist, but their use is generally limited to benign sea-state conditions. In commercial practice, when ultralarge container ships cannot enter a port because their draft exceeds the controlling depth of a harbor, a sheltered area is selected outside of the harbor and containers are transferred to locally available lighters or barges.

Although commercial crane technology is highly developed, limitations exist in its use in skin-to-skin transfers. Ships with widely different hull shapes or superstructure designs cannot come alongside each other in high-sea or swell states even when significant fenderage capability is available. In some situations, a two-step process is used to solve the problem of ship-to-ship transfers between incompatible hull forms. A large, heavily fendered barge is placed between the two incompatible hulls, and the cargo is transferred from one vessel to the barge and then from the barge to the second vessel.

The problems of crane design for open-ocean skin-to-skin transfers are substantially mitigated if either or both vessels involved have low pitch, roll, and heave motions as a result of their displacement (ballasted draft) and/or an active roll-mitigation system.

The cranes used for such purposes typically have a 75 ft reach for a 56,000 lb load and are 6 degrees-of-freedom (6-DOF) controlled platforms that actively compensate for movement of alongside platforms. Knuckle-booms with fixed-location pivoting bases are frequently employed. The design of the cranes incorporates active/passive compensation using hydraulic cylinders in the boom and in the knuckle and boom. The relative motion of the alongside platform is sensed, and predictions are made of future (short-term) platform motions so that the timing of cargo delivery can be optimized. In more sophisticated designs, dynamically tensioned guy wires are attached to the cargo in transit so that pendulous motions are mitigated.

The Office of Naval Research (ONR) has supported an S&T program in this area for a number of years. No programmatic roadmap exists for an end-item deliverable from this effort. The ONR should develop a programmatic roadmap for stabilized cranes for skin-to-skin transfers. The limiting factors appear to be the following:

- The need to develop an ability to provide a precise, short-term prediction of the motions of a receiving platform relative to the motions of the delivery platform, and
- The need to improve the response time of guy wire tensioning equipment designed to damp the pendulous motions of cargo being transferred.

Wire Line Transfers

The Navy has had a ship-to-ship wire line transfer capability for many decades. The capability to perform such transfers is limited to ships that carry the necessary equipment. In practice, this generally implies that wire line transfers can be accomplished between U.S. Navy (USN) ships and U.S. naval ships (USNSs; USNSs are civilian manned and in service), but not between arbitrarily selected commercial ships and USN/USNS ships. The committee presumes that any design for a sea base would incorporate the necessary capabilities for ship-to-sea-base wire line transfers.

Current Navy wire line transfers are limited to pallet loads of about 5,700 lb. Equipment limitations such as king post strength, cable strength, the response time of tensioning engines, and the torque of available hauling engines determine the limits to the weights that can be transferred.

A funded program exists to provide increased (about 14,000 lb) wire line weight-transfer capabilities in time for incorporation in the MPF(F) design. No technology problems are anticipated that would cause significant delay in the development of this capability.

Although a more robust design for a wire line transfer capability can probably be available as early as FY 2008, this enhanced capability is unlikely to be retrofitted into all operational USN/USNS vessels in fewer than 12 years. Thus, a full operational capability (FOC) for an enhanced wire line transfer capability is unlikely to occur prior to 2020.

Although wire line transfer capabilities exist on current USN/USNS vessels, it is not clear that all of the connectors being considered for use with the sea base will have a wire line connector capability. Certainly, current LCACs do not have such a capability, nor do the current TSV/HSV designs. Cargo will be transferred to such connectors by other means.

Liquid Transfer

Ship-to-ship transfer of fuel and in special cases other liquids has been accomplished routinely in both commercial and naval practice for many decades. Deep-draft commercial tankers normally are required to off-load some of their cargo to barges or transfer shuttles before entering harbors with limited controlling depths. Although this is not a standard Navy practice, the availability of fuel-transfer connectors might be an important adjunct to the sea base concept of operations.

The Navy's capability to accomplish underway refueling is excellent, and as far as is known, no S&T or R&D effort exists to improve the Navy's open-ocean fuel-transfer capabilities. Ship-to-shore fuel- and water-transfer requirements have not been defined and appear to be a limiting factor in some scenarios.

The major deficiency in liquid-transfer capabilities resides in the area of direct ship-to-shore transfer of liquids such as water. The Joint Logistics Over the Shore (JLOTS) program has dealt with the problem. The operational equipment needed for JLOTS liquid transfers is limited to a 3 to 4 mile range. No R&D program exists that is designed to produce equipment that will enable a greater standoff delivery range. As currently conceived, JLOTS will not be applicable to the sea base concept of operations unless the current JLOTS capability is incorporated into the design of a system that allows the routine transfer of liquids from a sea base to the shore connector. To the committee's knowledge, no program in support of the development of such a capability exists.

The sea-base-to-shore transfer of water is not nearly as much of a problem as the transfer of fuels. Water can be transferred to forces ashore by vertical lift either in bladders or as shrink-wrapped pallets of plastic water bottles. Furthermore, mobile ground forces generally have some autonomous water purification capabilities (although such capabilities might be of limited value in a desert environment).

Roll-on/Roll-off Ramps

Roll-on/roll-off (RO/RO) ramps are in widespread naval and commercial use. The Navy's large, medium-speed roll on/roll off (LMSR)-size ships have both stern and side ramps that allow wheeled and tracked vehicles to be driven from the LMSR. Current ramp designs are limited to pierside or causeway employment.

The use of RO/RO vessels is generally limited to major ports with a substantial pierside infrastructure that can accommodate deep-draft vessels. The number of such ports in the world is small relative to the total number of ports that might exist in future areas of military operations. In consideration of this limitation, the U.S. Army has maintained a program leading to the development of a deployable, floating causeway that can mate up with a deep-draft RO/RO ship. The problems involved in the development of such a capability are formidable. If the Army's development of this capability is successful, the value of RO/RO vessels as sea base connectors would be greatly enhanced, since their utility would not be limited to a relatively small number of ports.

Although the concept of roll-on/roll-off ramps is being considered for possible future inclusion in the MPF(F), a decision has been made to defer engineering studies of ramp configurations until 2008. In effect, this decision means that RO/RO capabilities will not be incorporated in the first flight of the MPF(F).

To some degree, the issue discussed above highlights a relatively fundamental difference between an MPF(F) and a sea base. Although a sea base could be loaded and unloaded alongside a pier or causeway, its primary design criteria will be an ability to transfer cargo to and from connector and resupply ships in the open ocean. Although the MPF(F) ships will certainly have a capability to transfer cargo to and from connectors and resupply ships in the open ocean, the first flight of the MPF(F) will be a relatively conventional vessel, possessing some open-ocean, cargo-transfer capabilities, that will be most easily unloaded while alongside a pier and causeway. Considerations are being given to future modifications of the MPF(F) for incorporating additional features within the design that will make at-sea cargo transfer more efficient over a wide range of sea states. A decision to configure the MPF(F) vessel in this way constitutes a decision to configure the sea base at what was termed Level One in Chapter 1 of this report—that is, a modest improvement in current capability, with little potential for tactical support for forces ashore from the sea.

Several concepts have been put forward for the use of a ramp with a sea base. The ramp from the sea base might be designed to terminate on an ILP or on a MLP. A short bridge might connect either the ILP or MLP with a TSV/HSV, an LCAC or an B-HSC. No such designs have been adopted. Other than conventional weight/moment and ramp-load capacity decisions that ultimately must be made, there are no currently perceived problems that would preclude the incorporation of such capabilities into an MPF(F) design. However, one must recognize that ramp locations and functions are major drivers of ship designs. They affect virtually the entire internal arrangement of the ship as well as its structural design. In the case of a stern ramp, for example, if one is to be fitted, the internal deck arrangements must permit it to be stowed. Ramps cannot be easily retrofitted into an existing design if the ship class was not designed from the start with the ramp locations and functions identified.

If a decision is made to commence engineering design studies for a RO/RO ramp for MPF(F) in 2008, the committee is confident that such a capability could be in service between 2013 and 2016. In any case, it appears to the committee that any RO/RO ramp design that is adopted for a sea base should be part of an integrated design of the sea base and all associated connector vessels.

The key issues in the sea base context are the following:

- The locations and functions of RO/RO ramps will drive the designs of the core sea base ships as well as the designs of high-speed connectors, and
- If the ramps are to be used for open-ocean cargo transfers, they must be tested at sea in realistic environments prior to committing the sea base and connector ships designs to them.

Integrated Landing Platforms

An integrated landing platform may be defined as a large floating platform (or raft) with reserve buoyancy of at least 200 to 250 tons. This buoyancy requirement is established by the need to support the weight of an LCAC and its payload.

The ILP concept of operations (CONOPS) is that an LCAC will drive itself onto the ILP that will be contiguous to an MPF(F) or to the sea base. Because of the lack of significant relative motion between the ILP and the mother ship, the transfer of loads to the LCAC will be simplified.

Although the capability to service an LCAC from an experimental ILP has been demonstrated, the final configuration of an ILP design that will be compatible with an MPF(F) hull has not been decided upon. The ILP will probably weigh about 200 to 250 tons and have dimensions of at least 100 ft by about 60 ft (the footprint of an LCAC is 87.5 ft (26.8 m) × 47 ft (14.3 m); the LCAC-X will be 1.66 times as long as the LCAC). Deploying an object of this magnitude and stowing it for transit will require the development of special handling equipment and the appropriate allocation of reserve weight and moment.

Current efforts on the development of the ILP are oriented toward its integration into the first flight of the MPF(F). The committee believes that a number of engineering design issues must be resolved prior to the incorporation of an ILP into the MPF(F). The committee does not believe that the resolution of these issues will require any S&T investment.

Realistic, full-scale, at-sea testing of the ILP and its connecting ramps must be carried out in order to confirm that they work successfully and that the required cargo transfer rates can be achieved. Also, the committee believes that this testing must be accomplished before committing the MPF(F) design to the ILP, so that an alternative transfer method could be adopted if required.

If studies show that the availability of an ILP will provide a significant enhancement of the cargo discharge rate of an MPF(F), ILPs are likely to be incorporated into the first ships of this class and might well be available operationally by 2012 or 2013.

Mobile Landing Platforms/Float-on/Float-off Heavy-Lift Ships

For several decades, the Navy and Marine Corps have had a capability of ship-to-shore transfer of cargo and equipment using landing craft, air cushion. LCACs ride on a high-pressure air cushion and can transport militarily significant payloads at relatively higher speeds than are normally achieved with conventionally designed monohulls.

One pays a price for these capabilities in that the Von Karman transport efficiency of an LCAC is low (ton-per-mile costs are high because fuel consumption is high). As a consequence, the payload range characteristics of LCACs are not compatible with the sea base CONOPS that constrain the sea base to operate at distances of greater than a hundred miles offshore. No extant LCAC design provides for round-trip ranges of these magnitudes.

Because of the attractiveness of an LCAC as a transport system that can deliver heavy equipment directly onto the beach, people have sought designs for LCAC shuttle ships that can transport fully loaded LCACs from the offshore sanctuary distances of a sea base to ranges from a hostile shoreline.

Heavy-lift ships (HLSs) have been used successfully in commercial and military salvage operations for many years. Basically, an HLS is a vessel with a large, strong, unencumbered flat deck that can be ballasted down so that its deck is awash or even below the surface of the ocean. In operation, the vessels that are to be transported flow onto the awash or submerged deck, and the HLS is then deballasted. Conceptually, the HLS would then transport the LCAC to a location where it would be capable of operating autonomously. When the LCAC had delivered its cargo to the beach, the process would be reversed and the HLS or mobile landing platform would return the LCAC to the vicinity of the sea base.

Since float-on/float-off HLSs exist, technology is not a limitation. There are shipbuilders who would be happy to provide a response to a request for a proposal. If the Navy decided to procure an HLS, it could have an operational capability in 5 to 6 years (the normal time to build a new ship design). In the past, if the Navy needed a float-on/float-off capability for a specific salvage project, it rented the services of a commercial HLS.

Although the HLS was not designed to deliver cargo, it can be used as an intermodal transfer platform. Studies of the rate of delivery of equipment to shore using such a system are not available. Since the hull forms of current HLSs are not designed for high speed, the overall equipment-delivery capability of the MLP concept may be unacceptably low unless current designs are modified to incorporate higher-speed operation.

As will probably be the case, the slow speed of the HLS may preclude its operational employment as an LCAC shuttle from sea base standoff ranges greater than 100 nmi. In that case, a high-speed LCAC shuttle would be needed with a sustained speed greater than 30 knots. This could be done with a monohull or with other hull configurations. However, the committee notes that the HLS has another critical role in the sea base—that of a transfer platform, used to facilitate vehicle transfers by RO/RO between supply ships and the sea base and also in the transfer of cargoes between the MPF(F) ships and the HSCs.

Well Deck Loaded Air Cushion Vehicles

For several decades the Navy's amphibious ships have had an operational capability to transfer cargo to LCACs located within the ships' flooded well deck enclosures. As sea states increase, so do the pitch, roll, and heave motions of the ship. These motions, if severe enough, will cause sloshing motions in the water of the well deck, which in turn eventually force a suspension of the loading of the LCAC.

Although the well deck can be pumped dry to eliminate the effects of water sloshing, the high-sea-state-induced pendulous motions of cargo suspended from the ship's internal gantry will force termination of operations. In principle, the technology of stabilized cranes and dynamic guying systems that are employed for skin-to-skin cargo transfers could be used with the internal gantries of well deck amphibious ships. At present, there appears to be no program whose goal is to produce a stabilized internal gantry within amphibious ships.

The principal limitation of the use of well decks to load LCACs is the large internal cubage of the mother ship that is preempted by a well deck and the limited number of discharge points available on a well deck amphibious vessel.

Transverse (athwart ship) dry wells might be considered and experimented with. A transverse dry well configuration would reduce the ship impacts of a well and permit one-way "drive through" LCAC transits with loading from above. The well closures at the hull side could be dropped to form the entrance and exit ramps for LCAC transit. In higher sea states, the doors on the weather side of the well could be closed, and the LCAC could enter and leave (by backing) on the lee side of the MPF(F) ship. LCAC loading would be done in a sheltered environment and could be done quickly by dropping large, preloaded pallets down from above. Cargo griping to the pallets (the most time-consuming part of the loading evolution) could be done in advance on the deck above. Cargo flow rates could be significantly improved by these means.

As briefed to the committee, current thinking seems to be that well decks will not be included in the design of the MPF(F). The use of stabilized cranes and the use of ILP and MLP are believed to obviate the need for a well deck to allow transfer of cargo from an MPF(F) into an LCAC. If these three technologies are mature by the time of initial operational capability (IOC) of the MPF(F), then well decks will not be required to load an LCAC.

Clearly, the three technologies referred to here must be proven before the MPF(F) design is frozen, not by the IOC date. If the MPF(F) will require a well deck (longitudinal or transverse) or a stern elevator and/or a stern ramp, these features must be designed into the ship from the outset. They cannot be added to the ship after delivery if the ship has not been designed to accommodate them.

The remaining question will then be, how do the LCACs reach the sea base or the theater of operations? LCACs do not have a trans-ocean self-deployment capability. One possibility may be to use an MLP as a CONUS base (or intermediate support base (ISB)) to the sea base shuttle ship. Alternatively, the LCACs might be carried as deck cargo on the MPF(F), on the sea base ship, or on the high-speed support ship, which is conceptually designed to transport CH-53E helicopters (CH-53X in the future).

ANTICIPATED TIME HORIZONS FOR THE TECHNOLOGIES

Most of the technologies and capabilities that are not currently available to support the desired sea base cargo-transfer capabilities appear to represent little more than an ensemble of moderately difficult engineering design problems. The current S&T efforts in support of the development of new or improved cargo-transfer capabilities appear to be minimally funded and do not seem to have associated dates for completion of the efforts. To the extent that any time lines exist for either S&T or R&D efforts in this area, they appear to be keyed to the fielding of MPF(F) ships or to a retrofit into future flights of the MPF(F). This implies that some of the current R&D programs for improved high-sea-state cargo-transfer capabilities may appear in about 2012 or 2013 when the first MPF(F) is commissioned. Few members of the MPF(F) design community predict that improvements in high-sea-state cargo-transfer capabilities will result from current S&T investments prior to 2020 or 2025.

The committee believes that the development of enhanced high-sea-state cargo-handling capabilities will be achieved if and only if they are integrated into the ship design process. For example:

- The degree of ship-roll compensation incorporated into the MPF(F) vessel or sea base design may determine how difficult it will be to design a stabilized crane that can transfer a TEU in Sea State 4;
- The size, displacement, and number of ILPs carried by the sea base will strongly influence the design, number, and total cargo throughput of the chosen connector vessel; and
- The design of the flight deck and possible inclusion of an electromagnetic catapult will determine the feasibility of using STOL aircraft in lieu of develop-

ing a new family of heavy-lift VTOL. As indicated in Chapter 3, it may be necessary to separate the design of ships to handle heavy cargo by airlift and by seaborne connectors; this would lead to having two kinds of ship in the sea base, with attending operational complexities and cost.

Based on briefings that it received, the committee is not sanguine that any of the desired high-sea-state cargo-handling technologies for sea base will be available within 10 years. Although a few programs appear to exist whose end item deliverable will be available for inclusion in the first MPF(F), the MPF(F) design efforts do not appear to assume successful completion of any current S&T or R&D efforts. Briefers made vague allusions to future retrofits of new technology, but no thought appears to have been given to the implications of such technology on the overall design of the sea base, on the design of the sea-base-to-shore connector fleet, and on the requirements for the design of logistic vessels for resupply of the sea base.

TASKS FOR THE FUTURE: THE WAY AHEAD

The committee does not believe that additional studies are required at this point in order to achieve enhanced high-sea-state cargo-handling capabilities. What appears to be required is an integrated engineering development program. The committee has the strongly held view that high-sea-state cargo-handling capabilities must be integrated into the design of the ship or ships that will constitute the sea base. Near-term, full-scale, at-sea testing of cargo-transfer options, including the ILP, transverse dry well, stern elevator, and stern ramp to the HLS, should be completed, and the results should be assessed before the design of the MPF(F) is frozen.

All of the decisions regarding open-ocean cargo transfer should be informed by advance knowledge of the desired level of capability to be achieved in the sea base. Unless this determination is made in the near future, the elements of the sea base will be mismatched, and the overall level of capability will not be fully optimized.

The committee believes that a requirement for the cargo-transfer components of the sea base should be common (as opposed to being interoperable) across the Services. Although the case for a unified, systems-engineering-based management of all sea-base-related programs is made in Chapter 5 of this report, the need for such an approach is emphasized here because meaningful progress will not occur without it.

The committee believes that unless a large testbed (possibly an LMSR-size ship and later, perhaps, a reserve carrier activated for experimentation with fixed-wing, heavy-lift aircraft, if that proves necessary) is made available to test engineering designs for improved, high-sea-state cargo-handling concepts, new cargo-handling capabilities are unlikely to be fielded by the Navy within the next 10 to 20 years.

CONCLUSIONS AND RECOMMENDATIONS

If the United States is to attain a true sea base capability rather than a maritime prepositioning capability, significant improvements must be achieved with regard to capabilities to transfer cargo and personnel to and from a sea base in high-sea-state conditions. A high-sea-state, open-ocean, cargo-transfer capability will be essential because U.S. forces may encounter limitations to port access through some combination of the following:

- Nonexistence of ports;
- Shallow, limiting depth of ports;
- Limited diameter of turning basins;
- Docks and piers unable to sustain meaningful military loads (greater than 40,000 lb);
- Cruise and ballistic missile attacks on harbor facilities;
- Mined harbors and approach areas;
- Entry channels physically obstructed (by sunken ships); and
- Political denial.

Managing at-sea loading and off-loading of the sea base requires a complex mix of capabilities for dealing with prevalent sea-state conditions in the most likely littoral environments. While existing capabilities and technologies permit some at-sea transfers in nonport environments during a wide range of sea-state conditions, the implementation of a joint sea basing for force projection necessitates a coordinated development and testing of open-ocean loading and off-loading capabilities for large cargo packages in conditions as severe as Sea States 3 and 4.

Most sea base concepts are based on the assumption that an intermediate support base will be available. Such ISBs may not be "sovereign" and might themselves become political, terrorist, or military targets for an adversary. Sea base design concepts should be based on the technical implications of the possible nonavailability of an ISB. In the committee's view, this consideration further strengthens the need for a sea base to be able to accommodate heavier VTOL aircraft (e.g., C-130-like or hybrid-lifting bodies) and to accommodate heavy-lift, intership transfers.

Even if intermediate bases are available and within a reasonable distance from the intended theater of operations, the United States will need to develop and maintain a capability to deliver personnel and materiel from a sea base to the shore. This implies that there is a need to invest in technology (R&D) for the following:

- Heavy-lift aircraft (about 40,000 to 43,000 lb) that can land and take off from a sea base in super-short-takeoff-and-landing (SSTOL), short-takeoff-and-vertical-landing (STOVL), and possibly in full VTOL mode;
- A capability that will provide for the rapid upgrade of marginal ports and their infrastructure;
- Heavy-duty causeways that can be deployed rapidly and are compatible with RO/RO ships; and
- A capability that will allow open-ocean cargo transfer in high sea states from a sea base to high-speed surface connectors/litters (possibly with beachability), and/or LCACs with improved performance.

Regarding the capability for open-ocean, heavy-cargo handling, the committee concludes the following:

- Attainment of high-sea-state cargo-handling capability will not be achieved until such capabilities are integrated into a ship design process that includes the associated connectors;
- Presentations that it received did not convince the committee that either a sense of urgency or the required funding support was attached to developing the desired high-sea-state cargo-handling technologies; and
- Absent a large-displacement (possibly an LMSR-size ship), joint testbed vessel to experiment with engineering designs and prototypes—or possibly two vessels, to experiment with heavy-lift aircraft and ship-to-ship cargo transfer if it is found that the capabilities cannot be combined in a single ship—the committee believes that enhanced high-sea-state cargo-handling capabilities may not be available for fielding by the U.S. Navy in the targeted timeframe.

The committee recommends the following:

Recommendation: The Department of the Navy should accelerate efforts to achieve (1) a capability for skin-to-skin transfer of cargo, in Sea State 4 conditions, to and from a sea base and arbitrary commercial cargo ship design, and (2) improved capabilities to transfer military cargo from a sea base to the high-speed surface connectors that move cargo from the sea base to shore.

Recommendation: The Department of the Navy should identify one large vessel to be used as a testbed for resolving the known problems, including those related to connectors and internal cargo handling, involved in at-sea cargo transfer at Sea States 3 and 4, or two such vessels if required for an integral flight deck in order to explore issues associated with potential future heavy-lift aircraft.

5

A System-of-Systems Engineering Approach for Joint Sea Basing

BACKGROUND

Previous chapters in the report documented the specific conditions that have demonstrated that the joint Sea Basing concept lacks a clear joint definition and is currently characterized by disparate organizational and institutional views. The lack of an approved overall joint Sea Basing vision and empowered centralized planning authority has resulted in divergent efforts, a lack of clear communication among and within the military Services, and the absence of an effective top-level process or mechanism that might be successful in identifying and enabling a coordinated joint path forward.

While the committee agrees with the conclusion of the Defense Science Board Task Force on Sea Basing that Sea Basing is a critical, national, joint military competency to project forces rapidly from the United States, as yet no effective organizational entity has been commissioned that will bear the task of bringing this future joint military capability to reality.[1] Even the basic working definitions and therefore the candidate technical solutions for achieving a joint Sea Basing capability differ by Service, resulting in little unanimity on key aspects of the way ahead in fielding a joint sea basing capability.

The crux of the problem, as determined by this study, is that one cannot realistically expect the Services to individually or collectively design the grand

[1] Defense Science Board. 2003. *Defense Science Board Task Force on Sea Basing*. Office of the Under Secretary of Defense for Acquisition, Technology, and Logistics, Washington, D.C., August.

scheme of joint integrated Sea Basing absent increased top-down guidance and a centrally managed system-of-systems approach.

The complexity and difficulty of transforming diverse aspects of extant and programmed U.S. military capabilities in order to achieve an effective future joint Sea Basing competency require the coordinated development of a joint Sea Basing system of systems. The successful development of joint Sea Basing capabilities for the nation involves long-term capital investments in interdependent complex platforms, connectors, and supporting technologies, some of which exist today or are under development and some of which are yet to be developed. To succeed, such an effort must of necessity involve all of the military Services and the appropriate U.S. government agencies in a coordinated development process under a coherent set of ultimate goals, common standards, integrated requirements, and time-phased, mutually-accepted priorities.

MAIN ELEMENTS OF A SYSTEM-OF-SYSTEMS APPROACH

Creation of a Top-Level Joint Sea Basing Concept Vision

Briefings to the committee indicated that there currently are widely divergent views among the various stakeholders and Service representatives regarding the joint Sea Basing mission, concept of operations, and required capabilities. This tendency to see the concept from primarily single-Service perspectives can be expected to continue until an empowered senior central authority exists to generate an approved, official Joint Sea Basing Vision. Given the complexity of a joint Sea Basing system-of-systems approach and the long-term nature of the major capital investments by Services in new platforms, development of advanced technologies, and the introduction of appropriate joint doctrine, such a unifying vision will be essential in order to best leverage existing currently programmed and future Service capabilities. The office that creates the Joint Sea Basing Vision will also have a continuing role in coordinating the long-term spiral development of the joint Sea Basing capability. Key provisions of the overarching Joint Sea Basing Vision should include the following:

- The joint Sea Basing mission statement,
- The joint Sea Basing concept of operations and employment,
- Identification of specific Service and agency roles and missions within the joint Sea Basing concept,
- The designation of a Joint Sea Base Planning Office that can grow into a Joint Program Office for the Sea Base and the appointment of a flag or general officer as the responsible system-of-systems executive to implement the Joint Sea Basing Vision and manage joint Sea Basing spiral development,

- The designation of a lead Service or Services to conduct joint Sea Basing field testing and experimentation,
- The appointment of a lead Service or lead agency for joint Sea Basing joint doctrine development,
- The publication of a joint Sea Basing technology roadmap, and
- The publication of a joint Sea Basing plan of objectives and milestones (system-of-systems master time line).

Include All Service and Government Agency Programs in the Joint System-of-Systems Management Plan

The designated Joint Sea Base Planning Office should centrally manage and deconflict requirements-generation and procurement programs for all Service components participating in the implementation of joint Sea Basing. The purpose of this effort would be to correlate Service requirements and advise Service procurements to appropriately exploit the interdependence of Service-specific capabilities while avoiding excessive capability duplication, redundancy, and conflicting acquisitions. This system is not intended to replace the current Joint Capabilities Integration and Development System (JCIDS) or acquisition process but to provide a single source that coordinates and advises individual Services as inputs to the current joint process.

Coordinate the Development and Testing of All Relevant Sea Basing Technologies

Given the complexity and diversity and long-term impact of individual Service's technological development and implementation on closely related programs, such as the sea base surface and air connectors with the larger joint prepositioning ships, a centralized management and coordination agent is necessary to correlate, deconflict, and integrate technology development according to the Sea Basing master plan. Only a central agency can possibly optimize technological synergies, maximize multi-Service applications, and minimize unnecessary redundancy or the duplication of currently uncoordinated efforts.

Identify the Interrelationships Among Decisions on Diverse Programs, Technology, and Force Structure

It was clear to the committee that individual Service decisions on specific technology roadmaps, force structure options, and developmental decisions and doctrines would have far-reaching impacts, including secondary- and tertiary-level ramifications, within and across broader Service and joint capabilities, doctrine, and force structure. One of the primary functions of a Joint Sea Base

Planning Office would be the identification and management of the long-term ripple effect of various decisions and technology paths chosen. For example, the composition of the future Expeditionary Strike Group (ESG), with either two or three large-deck amphibious ships, would have great impact on the capability gap analysis in terms of determining what future capability must be included in future programs. The doctrinal decision as to whether a Maritime Prepositioning Force (Future) (MPF(F)) squadron would be expected to operate in a hostile environment absent the Sea Shield umbrella provided by Carrier Strike Groups (CSGs) and ESGs would have great impact on self-defense and survivability requirements and therefore on the cost of any future MPF(F) ship design.

There are also questions relating to force structure that center on two main issues: command and use. If the sea base ships come under the Military Sealift Command (MSC), the Navy, the Joint Chiefs of Staff, and so on, would somewhat lose control of these ships, as they would be manned by civilian crews and masters. Thus, unless specific agreements were made up-front, the degree of military control would be constrained unless the ships were involved in actual combat. If the ships are intended to be built to "combat" rather than to "commercial" survivability standards, will they be crewed by U.S. Navy personnel? If so, many force structure implications that impinge on recruiting, training, retention, troop strength, and so on must be addressed.

SYSTEM-WIDE ISSUES FOR THE SEA BASING CONCEPT

Strategic Airlift and Sealift

Strategic airlift and sealift will continue to play an important role in ferrying troops from the continental United States (CONUS) to the sea base and in surging heavy equipment and resupply cargo from CONUS to advanced bases, intermediate bases, and, if possible, the sea base. There is much to be done in working with the Air Mobility Command (AMC) and the Military Sealift Command in identifying specific lift requirements and resources that might be available in a wide variety of national security contingency scenarios. Regarding strategic airlift and sealift, the committee concludes the following:

- It is not clear that the sea Services have adequately assessed future capabilities and interdependencies between AMC airlift and MSC sealift across the spectrum of potential scenarios for future crises and conflicts. Nor has any analysis been conducted to include the increased capabilities inherent in intratheater systems such as the Navy's future littoral combat ship and the Army's theater support vessel platforms.
- Increased cross-pollination with the AMC and the MSC and the Services is required in order to develop an integrated path forward in providing the best

balance in the strategic and operational lift requirements for joint Sea Basing operations. In addition, theater support airlift, with precision airdrop capabilities or the possibility of landing (but not refueling) at forward locations, will also be available, if needed, to sustain the forces ashore as appropriate and feasible.

Command, Control, Communications, Computers, Intelligence, Surveillance, and Reconnaissance Requirements

Establishing, defending, and operating the joint sea base in either a contested or benign littoral environment will require a complicated, networked architecture of joint command and control, active and passive situational awareness, information fusion, and the ability to pass these capabilities easily from the sea base to operations ashore and then back again.

The key questions regarding command, control, communications, computers, intelligence, surveillance, and reconnaissance (C4ISR) requirements are as follows:

- Who will provide the bulk of this C4ISR support? Where will it be located?
- What is the joint C4ISR concept of operations and where is the joint master plan for bringing these capabilities online?

Sense-Respond Logistics and Improved Inventory Control

Most of the presentations before the committee focused on the hardware, platform, and connector facets of the joint Sea Basing concept. There needs to be equal emphasis on the new technologies and processes required for enabling efficient, sense-respond logistics concepts; conducting effective, selective on- and off-load operations; and effecting timely reload and redeployment options.

Essential Technologies Roadmap

Briefings to the committee convinced its members that there are many useful and potentially transformational technologies under development. However, the degree of compartmentalization of efforts, the absence of an empowered cross-Service coordination venue, and the lack of an overall Sea Basing technology roadmap provided a compelling argument for establishing a system-of-systems approach. Regarding the needed Essential Technologies Roadmap, the committee concludes the following:

- There is agreement on some of the fundamental essential enabling technologies for attaining the joint Sea Basing concept. However, there is currently no prioritized technology-development agenda and no centrally maintained time line for tracking the parallel development of relevant technologies that might be accelerated or adapted to multimission/multi-Service use.

- Only the publication of an approved Joint Sea Basing Essential Technologies Roadmap can bring order and synergy to the current, largely unstructured, and often divergent approach to technology development.

VALUE OF A SYSTEM-OF-SYSTEMS APPROACH TO ACHIEVING JOINT SEA BASING CAPABILITY

A system-of-systems approach to achieving joint Sea Basing capability would be of value in the following ways:

- As "conceived-and-born-joint" capability—the best path to achieving a joint Sea Basing vision and concept of operations;
- For avoiding excess costs of implementation by helping to prioritize individual Service and agency investments;
- By offering incentives for joint contributions and investments;
- Through the identification and coordination of parallel or sequenced development; and
- As the best structure for identifying and anticipating the ripple effect of early decisions upon other related components.

RECOMMENDATIONS

Recommendation: A Joint Sea Base Planning Office—directed by a Navy flag officer or a Marine Corps general officer—should be established. The director of this office should report to an appropriate official in the Office of the Secretary of Defense.

The Department of the Navy widely describes Sea Basing as being transformational. The committee agrees with this description. If implemented as described in the concept document,[2] Sea Basing will occasion fundamental changes in the way that the Navy and Marine Corps operate, and it will create a truly new National Joint Capability for the future. Sea Basing would have a profound effect on force structure, logistics, training, and supporting infrastructure. Thus, it cries out for a management system that can emphasize and implement an effective systems engineering approach. The current system—with concept development, research and development, procurement, modeling and simulation, and other

[2]ADM Vern Clark, USN, Chief of Naval Operations; and Gen Michael W. Hagee, USMC, Commandant of the Marine Corps. 2003. *Naval Operating Concept for Joint Operations,* Department of Defense, Washington, D.C., September 22.

activities all apparently proceeding independently, and in some cases redundantly or on diverging tracks—will not yield the envisioned capability.

Recommendation: In order to facilitate the management of multi-Service efforts through the use of common terminology and technologies as they create the various elements of the sea base and the airborne and seaborne connectors, and to focus the overall effort, the entire sea base effort within the Joint Sea Base Planning Office should be managed under a name—for example: "Joint Maritime Prepositioning and Sea Basing Force."

The committee recommends putting all of the various potential ship and aircraft types and programmatics, whether near term, long term, or thus far only in concept development, under a single rubric and management so that the future prepositioning force can be centrally managed for interoperability, complementary capabilities, and maximum overall efficiency, affordability, and warfighting synergy. Such an organizational construct would facilitate the orderly incorporation and spiral development of standardized and complementary new entries and improved designs over the long life of the Joint Maritime Prepositioning and Sea Basing Force (Future) effort.

Recommendation: The U.S. Joint Forces Command should be designated along with the U.S. Navy and U.S. Marine Corps as the initial lead Services in managing and executing the joint Sea Basing spiral development process.

Recognizing that while full jointness in Sea Basing remains the ultimate goal, suitable force structure, existing or near-term capability, and relevant past expertise in this realm today reside almost exclusively within the Navy and the Marine Corps. The Army plans for a new Combat Prepositioning Force, to include the Afloat Forward Staging Base, are evolving within Army transformation and are not yet at a stage of maturity but should be included. Under the current approach to Sea Basing, the Air Force has been largely ignored, but it should instead be included as a major participant in the primary area of force deployment and sustainment.

Recommendation: The Joint Strike Fighter program should be used as a model to manage the joint Sea Basing spiral development process.

The Services' recent experience with developing, coordinating, and managing the immensely complicated Joint Strike Fighter program provides a suitable model for effecting the broader transformational process; this process could bring diverse Service and agency force structures, cultures, and capabilities to a new level of cooperation and interdependence as embodied in the joint Sea Basing concept.

Recommendation: The Joint Sea Base Planning Office should create a joint master plan for technology development, based on an integrated systems-of-systems engineering approach, for the Services to use in developing the components of the sea base within their assigned jurisdictions.

As stated above, the early publication of a comprehensive technical roadmap is the first step in introducing a system-of-systems approach to Sea Basing and would go a long way in bringing order and purpose to the disparate array of independent research and development vectors that characterize today's Sea Basing development efforts.

Recommendation: A "platform-agnostic" joint capabilities-based approach should be applied to developing the initial joint Sea Basing doctrine, procedures, operational concepts, network-centric architectures, and mission requirements.

Resolution of many complex and difficult decisions regarding specific platform, connector, and hardware choices will be shaped over a lengthy period of joint experimentation and the joint spiral development process. To jump-start the joint Sea Basing development process, it would be better to concentrate initially on developing a platform-agnostic approach, which focuses on a system-of-systems look at appropriate Service roles and missions, new requirements, tactics, doctrine, processes, essential technologies, and desired capabilities.

Recommendation: The Department of the Navy should identify one large vessel to be used as a testbed for resolving the known problems, including those related to connectors and internal cargo handling, involved in at-sea cargo transfer at Sea States 3 and 4, or two such vessels if required for an integral flight deck in order to explore issues associated with potential future heavy-lift aircraft.

A joint testbed platform, or two platforms of different types if necessitated by the heavy-lift aircraft issue, should be used by the Department of the Navy for joint experimentation and spiral development of the joint Sea Basing concept/ Maritime Prepositioning Force (Future) capabilities and prototypes.

The committee recommends designating a large-displacement vessel as the Sea Basing joint testbed for the conduct of joint experiments to evaluate hull design modifications, connector interfaces, advanced at-sea cargo-handling technologies, and relevant operational, logistics, and information technology initiatives and innovations. Joint Sea Basing spiral development will clearly proceed as an ongoing effort of experimentation and prototyping well beyond initial operational capability as Service doctrine and requirements are refined. A well-coordinated and appropriate joint experimentation and prototyping investment is required in order to successfully and efficiently integrate and field a new joint Sea Basing capability that will be relevant in answering the nation's future needs.

Some large vessels, such as an Army large, medium-speed roll on/roll off and, if needed, a reserve carrier, may already be available in the inventory being maintained as contingency assets for rapid activation in a crisis and should be considered as suitable Sea Basing testbeds.

Appendixes

A

Committee and Staff Biographies

Harry W. Jenkins, Jr., retired from the U.S. Marine Corps with the rank of major general. He is the director of business development and congressional liaison at ITT Industries, where he is responsible for activities in support of tactical communications systems and airborne electronic warfare systems with the Navy, Marine Corps, Coast Guard, National Guard, and appropriate committees in Congress. General Jenkins's background is in expeditionary warfare, particularly in regard to its mission use of command, control, communications, computers, and intelligence (C4I) systems. During Desert Storm, General Jenkins served as commanding general of the Fourth Marine Expeditionary Brigade, for which he directed operational planning, training, and employment of the ground units, aviation assets, and command-and-control systems in the 17,000-person amphibious force. General Jenkins's last position before his retirement from the U.S. Marine Corps was as director of expeditionary warfare for the Chief of Naval Operations. In that position he initiated a detailed program for C4I systems improvements for large-deck amphibious ships, as well as managing all programs of naval mine warfare and reorganizing the Navy's unmanned aerial vehicle efforts for operations from aircraft carriers and amphibious ships. He is a member of numerous professional societies, including the Marine Corps Association, Marine Corps Aviation Association, Expeditionary Warfare Division of the Naval Defense Industry Association, Navy League, and Adjutant Generals Association of the United States. General Jenkins is a member of the Naval Studies Board.

Richard L. Wade is a principal at Exponent; formerly, he was president of Risk Management Sciences, a private consulting firm specializing in risk management and threat assessment. In addition, he is an adjunct associate professor of medicine at the University of California at San Francisco Medical Center. His areas of expertise include the commercial shipping industry, risk mitigation, threat assessment, and environmental and public health issues. Dr. Wade has served as the head of public health agencies (in Seattle, Washington, and for the states of Minnesota and California). In 1990, he received the American Public Health Association's lifetime achievement award. Dr. Wade is a member of the National Research Council's Naval Studies Board.

Jeffery P. Bennett is vice president of logistics management at LMI where he oversees studies and analysis for the Office of the Secretary of Defense as well as the individual military Services, for the Department of Homeland Security, and for other federal government agencies. His expertise is in supply chain management, cost and resource analysis, strategic logistics, operations research, and in naval logistics and underway replenishment operations. Mr. Bennett spent much of his career in Department of Defense (DOD) programming and budgeting organizations; he was a member of the Senior Executive Service and served as director of the Force and Infrastructure Cost Analysis Division (FICAD) for the Office of the Secretary of Defense, Program Analysis and Evaluation. Mr. Bennett led FICAD studies and analysis supporting independent cost estimates for the DOD Cost and Improvement Group, defense-wide budget and program reviews, and the first Quadrennial Defense Review. He earned his commission in the Navy through the Navy Enlisted Scientific and Education Program, graduating from the University of Washington with a B.S. in mathematics and later receiving his M.S. (with distinction) in operations research from the Naval Postgraduate School, accompanied by the Military Operations Research Society Graduate Research Award.

Alan Berman is an independent consultant, with current clients including the Applied Research Laboratory of Pennsylvania State University (ARL/PSU) and the Center for Naval Analyses (CNA). Dr. Berman's expertise includes analyses of Navy research and development investments, space operations capabilities, information operations, and command, control, communications, computers, intelligence, surveillance, and reconnaissance (C4ISR) programs. Dr. Berman served as dean of the Rosenstiel School of Marine and Atmospheric Sciences at the University of Miami, where he was responsible for the graduate programs in physical oceanography, marine biology, geology, geophysics, applied ocean science, and underwater acoustics. He also served as director of research at the Naval Research Laboratory, where he administered broad programs in basic and applied research. Dr. Berman has served on numerous scientific boards and advisory committees, including the recent Defense Science Board Task Force on Sea Basing.

Paul Bevilaqua is chief engineer of advanced development projects for Skunk Works at Lockheed Martin Aeronautics Company. He joined Lockheed Martin as chief aeronautical scientist and has spent much of his career developing vertical-takeoff-and-landing (VTOL) aircraft as well as playing a leading role in creating the Joint Strike Fighter program. He invented the lift fan propulsion system that made it possible to build stealthy, supersonic VTOL aircraft, and led the engineering team that demonstrated the feasibility of building variants of this aircraft for the Air Force, Marine Corps, and Royal Navy. Prior to joining Lockheed Martin, Dr. Bevilaqua was manager of advanced programs at Rockwell International's Navy aircraft plant, where he led teams designing short-takeoff, vertical-lift interceptor aircraft and VTOL carrier-onboard-delivery aircraft. He served as deputy director of the Energy Conversion Laboratory at Wright Patterson Air Force Base. He has a B.S. in aerospace engineering from the University of Notre Dame and M.S. and Ph.D. degrees in aeronautics and astronautics from Purdue University.

E. Richard Diamond, Jr., is senior manager of strategic assessments at Raytheon Integrated Defense Systems. His background is in senior-level war-gaming and simulations management with the Joint Staff, Naval War College, North Atlantic Treaty Organization (NATO), and various defense agencies. Mr. Diamond joined Hughes Aircraft Company and then moved to Raytheon following a career in the U.S. Navy, from which he retired with the rank of captain. At sea, he served on frigates, destroyers, cruisers, and aircraft carriers. He commanded the frigate USS *Kirk* and the Aegis cruiser USS *Bunker Hill*, both forward homeported in Yokosuka, Japan. Ashore, he founded the Chief of Naval Operations' first Joint Operations and Doctrine Branch, headed the Strategic Concepts Branch, and initiated the post-Cold War strategic reviews, which produced the U.S. Navy's future vision statements, *. . . From the Sea* and *Forward . . . From the Sea*. Mr. Diamond is a graduate of the University of Dallas, Tulane University Graduate School, and the U.S. Army Command and General Staff College. He attended Harvard University's John F. Kennedy School of Government and Tuft University's Fletcher School of Law and Diplomacy as a federal executive fellow. Mr. Diamond is currently an adjunct fellow at the Center for Strategic and International Studies.

Steven W. Flohr retired from the U.S. Army with the rank of brigadier general. He is an independent consultant on defense, technical, and program management issues. General Flohr's background is in the testing and evaluation of military systems, including those for joint operations, in defense and other government agencies and for industry and foreign customers. His last active duty assignment was as commanding general at White Sands Missile Range, where he directed missile range operations and installation activities to provide testing and evaluation for the joint Services. He served on the National Research Council (NRC)

Committee on Assessment of Test Infrastructure Requirements to Support Operational Testing of Defense Directed Energy Systems. General Flohr received his B.S. in mechanical engineering from the University of Nebraska and his M.B.A. from the Florida Institute of Technology.

Wesley L. Harris is the Charles Stark Draper Professor and head of the Department of Aeronautics and Astronautics at the Massachusetts Institute of Technology (MIT). His expertise is in fluid mechanics; aerodynamics; unsteady, nonlinear aerodynamics, acoustics; lean manufacturing processes; and military logistics and sustainment. Dr. Harris's background also includes managing major national and international aeronautical and aviation programs and personnel in the executive branch of the federal government. Prior to coming to MIT, he served as associate administrator for aeronautics at the National Aeronautics and Space Administration and vice president and chief administrative officer of the University of Tennessee Space Institute. Dr. Harris earned a B.S. in aerospace engineering from the University of Virginia and an M.S. and a Ph.D. in aerospace and mechanical sciences from Princeton University. He is a member of the National Academy of Engineering.

George B. Harrison retired from the U.S. Air Force with the rank of major general. He is director of research operations at Georgia Tech Research Institute, where he oversees sponsored research in aerospace, transportation, electronic systems, sensors, electronic combat, signature technology, information technology, and electro-optical applications. His background is in military operations, particularly in regard to air operations for the joint Services. Specifically, he commanded U.S. Air Force, Army, and Navy and French and British forces enforcing post-Desert Storm sanctions against Iraq. General Harrison received his B.S. degree in general engineering and public policy from the U.S. Air Force Academy and an M.B.A. in industrial management from the University of Pennsylvania. He has participated in the program for senior executives in national and international security at Harvard University.

Kevin F. Kelly is the director of strategic planning and advanced programs at Northrop Grumman Ship Systems. He is a retired naval officer with a background in operations, engineering, command-and-control systems with open architecture considerations, and future ship designs with an emphasis on the Navy's Sea Basing concept. Mr. Kelly received his B.S. in mechanical engineering from the U.S. Naval Academy and had 25 years of commissioned and enlisted service before beginning his career with Northrop Grumman. He received an M.E. in mechanical engineering from Tulane University and an M.B.A. from William Carey College.

Ronald K. Kiss is president of the Webb Institute, a private, 4-year college offering Bachelor of Science degrees in naval architecture and marine engineering. Prior to joining the Webb Institute, he was vice president of SYNTEK, assisting the U.S. Navy on the joint Navy and Defense Advanced Research Projects Agency arsenal ship program and on the Navy's aircraft-carrier and surface-combatant programs. He served as deputy assistant secretary of the Navy for ship programs in the Office of the Assistant Secretary of the Navy for Research, Development, and Acquisition, and also as executive director of the Amphibious, Auxiliary, Mine, and Sealift Directorate at the Naval Sea Systems Command. Mr. Kiss spent nearly 20 years with the Maritime Administration, completing his service there as acting associate administrator for shipbuilding and ship operations. He holds a B.S. degree in naval architecture and marine engineering from the Webb Institute, an M.S. in naval architecture from the University of California, Berkeley, and has participated in a number of post-graduate programs at institutions including Harvard University and the Massachusetts Institute of Technology.

John B. LaPlante retired from the U.S. Navy with the rank of vice admiral. He is an independent consultant and former manager of Department of Defense Business Development at McDermott International, Inc., a worldwide energy services company. His background is in naval (and joint) military operations, particularly in regard to operational logistics. Before retiring from the Navy in 1996, Admiral LaPlante served as director for logistics, Joint Staff. His military experience also included assignments as commander of the Naval Logistics Command Pacific and head of the Amphibious Warfare Branch in the Office of the Chief of Naval Operations. During Desert Storm and Desert Shield, he commanded all amphibious forces in the Gulf region, a force of some 43 ships and 34,000 men and women.

Henry S. Marcus is professor of marine systems in the Ocean Engineering Department at the Massachusetts Institute of Technology. He is chairman of the graduate program in ocean systems management and also served as chairman of the earlier MIT program in shipping and shipbuilding management. Dr. Marcus's expertise is in marine systems, marine transportation, ocean systems management, shipping, and shipbuilding management. He held the title of Naval Sea Systems Command Professor of Ship Acquisition for a decade and served on numerous committees related to maritime transportation sponsored by the NRC. Dr. Marcus served as a member of the Marine Transportation System National Advisory Council and the Federal Transportation Advisory Group. His most recent book, *Intermodal Movement of Marine Containers*, deals with the impact of new technology on the marine industry—case studies on ports, ocean carriers, and railroads analyze how technology brings about changes in strategy and competitive structure within the industry. Dr. Marcus has a B.S. degree from the Webb Institute, two M.S. degrees from MIT, and a Ph.D. from Harvard Business School.

Irwin Mendelson, retired president of the Engineering Division of Pratt & Whitney, is a mechanical engineer by education and training. His background is primarily in commercial and military aircraft engine design, with earlier experience in submarine design. At Pratt & Whitney, Mr. Mendelson managed a staff of 8,000 and oversaw a budget of $900 million. He was responsible for the total operation of the division, including the design, development, and installation of all aircraft engine systems. At Electric Boat, he had been responsible for major mechanical systems and components, including propulsion, missile and torpedo launch, and steering and diving controls. Mr. Mendelson received his B.S. from the Polytechnic Institute of Brooklyn and his M.S. from the University of Connecticut.

William B. Morgan retired after having spent his professional career at the Naval Surface Warfare Center, Carderock Division (formerly the David Taylor Research Center). He retired as head of the Hydromechanics Directorate. His expertise is in naval architecture, with a specialty in naval hydrodynamics; ship and submarine design, including seakeeping, maneuverability, cavitation, and control; design and development of advanced marine propulsors of all types; prediction of full-scale performance from model tests and computations; advanced hydrodynamic testing techniques and facilities and advanced computational techniques; and validation and certification of these various testing and computational techniques. Dr. Morgan has authored and coauthored numerous papers and reports dealing with a wide variety of subjects in the propulsion area. A member of the National Academy of Engineering, he has served on numerous scientific boards and advisory committees and is currently a member of the Naval Studies Board.

John H. Moxley III is managing director at the North American Health Care Division, Korn/Ferry International. A member of the Institute of Medicine, his expertise is in military medical issues, health science policy, human rights, international health, academic health sciences center policy issues, and the administration of pharmaceutical corporate/industry enterprises, federal government agencies, and hospital and medical centers. Prior to joining Korn/Ferry, he held a number of senior positions in academia, government, and commercial industry, including that of dean of both the University of Maryland and the University of California (San Diego) Medical Schools, assistant secretary of defense for health affairs, and senior vice president at American Medical International. He has served on numerous scientific boards and advisory committees, including the American Hospital Association board of trustees, the California Medical Association, the American Medical Association, the National Fund for Medical Education, and the Henry M. Jackson Foundation for the Advancement of Military Medicine. Dr. Moxley is a member of the Naval Studies Board and the Board on Army Science and Technology.

Robert G. Sprigg retired from the U.S. Navy with the rank of rear admiral. He is currently director of advanced warfare concepts with General Dynamics/Bath Iron Works. His background is in naval aviation, ship handling, and military logistics via joint and coalition support. Admiral Sprigg commanded the USS *Camden* while it was deployed in Operation Desert Storm/Provide Comfort providing logistical support to eight ships of foreign navies. He has also commanded the USS *George Washington*, CVN (nuclear-powered aircraft carrier)-73, Carrier Group Two, and the Navy Warfare Development Command. Admiral Sprigg graduated from the U.S. Naval Academy with a B.S. in naval science and earned an M.S. in aeronautical engineering from the Naval Postgraduate School.

Jennifer P. Whitlock is the Boeing Company's lead design engineer for airplane configuration for the Boeing blended wing body (BWB) subsonic transport program. Her expertise is in aircraft design and its configuration on carriers, the requirements for military cargo and commercial freight, as well as commercial passenger accommodations. Ms. Whitlock has been working with the BWB team since she started at then-McDonnell-Douglas in 1996. In addition, she has worked on the high-speed civil transport, the advanced theater transport, and the sonic cruiser tanker. She holds a B.S. in aeronautical engineering from the University of Illinois and an M.S. from Stanford University.

Staff

Charles F. Draper is acting director of the National Research Council's (NRC's) Naval Studies Board. He joined the NRC in 1997 as program officer, then senior program officer, with the Naval Studies Board and in 2003 became associate director. During his tenure with the Naval Studies Board, Dr. Draper has served as the responsible staff officer on a wide range of topics aimed at helping the Department of the Navy with its scientific, technical, and strategic planning. His recent efforts include topics on network-centric operations, theater missile defense, mine warfare, and nonlethal weapons. Prior to joining the Naval Studies Board, he was the lead mechanical engineer at Sensytech, Inc. (formerly S.T. Research Corporation), where he provided technical and program management support for satellite Earth station and small-satellite design. He received his Ph.D. in mechanical engineering from Vanderbilt University in 1995; his doctoral research was conducted at the Naval Research Laboratory (NRL), where he used an atomic force microscope to measure the nanomechanical properties of thin-film materials. In parallel with his graduate student duties, Dr. Draper was a mechanical engineer with Geo-Centers, Inc., working on-site at NRL on the development of an underwater X-ray backscattering tomography system used for the nondestructive evaluation of U.S. Navy sonar domes on surface ships.

Arul Mozhi is senior program officer at the National Research Council's Naval Studies Board and served as senior program officer at the NRC's Board on Manufacturing and Engineering Design and National Materials Advisory Board. Prior to joining the NRC in 1999, Dr. Mozhi was senior scientist and program manager at UTRON, Inc., a high-tech company in the Washington, D.C., area, working on pulsed electrical and chemical energy technologies applied to materials processing. From 1989 to 1996, Dr. Mozhi was a senior engineer and task leader at Roy F. Weston, Inc., a leading environmental consulting company working on long-term nuclear materials behavior and systems engineering related to nuclear waste transport, storage, and disposal in support of the U.S. Department of Energy. Before 1989 he was a materials scientist at Marko Materials, Inc., a high-tech firm in the Boston area, working on rapidly solidified materials. He received his M.S. and Ph.D. degrees (the latter in 1986) in materials engineering from the Ohio State University and then served as a postdoctoral research associate there. He received his B.S. in metallurgical engineering from the Indian Institute of Technology in 1982.

B

Agenda for the Committee's Meeting

Keck Center of the National Academies
Washington, D.C.
September 7-10, 2004

Tuesday, September 7, 2004

Closed Session:
Committee Members and National Research Council (NRC) Staff Only

1230 CONVENE—OPENING REMARKS, COMPOSITION AND BALANCE DISCUSSION
Harry W. Jenkins, Jr., Committee Co-Chair
Richard L. Wade, Committee Co-Chair
Charles F. Draper, Acting Director, Naval Studies Board (NSB)
Dennis Chamot, Associate Executive Director, Division on Engineering and Physical Sciences, National Research Council

Data-Gathering Meeting Not Open to the Public:
Classified Discussion (Secret)

1400 U.S. ARMY THEATER SUPPORT VESSEL
Tracy Mitchell, Assistant Program Manager, Objective Theater Support Vessel, Program Executive Office for Combat Support and Combat Service Support, U.S. Army Tactical Army Command (TACOM)

1530 U.S. ARMY SEA BASING REQUIREMENTS
BG David A. Fastabend, USA, Director, Concept Development and Experimentation, Futures Center, U.S. Army Training and Doctrine Command (TRADOC)

1630 SEA BASING SURFACE CONNECTORS
CAPT Patricia M. Sudol, USN, Program Manager, Ships, Boats, and Craft, Program Executive Office for Ships, PMS-325, Naval Sea Systems Command (NAVSEA)

Closed Session: Committee Members and NRC Staff Only

1730 COMMITTEE DISCUSSION—RECAP OF DAY 1
Moderators:
Harry W. Jenkins, Jr., Committee Co-Chair
Richard L. Wade, Committee Co-Chair

1800 END SESSION

Wednesday, September 8, 2004

Closed Session: Committee Members and NRC Staff Only

0830 CONVENE—COMMITTEE DISCUSSION
Harry W. Jenkins, Jr., Committee Co-Chair
Richard L. Wade, Committee Co-Chair
Arul Mozhi, Senior Program Officer, Naval Studies Board

Data-Gathering Meeting Not Open to the Public:
Classified Discussion (Secret)

0900 U.S. NAVY SEA BASING ANALYSES—OVERVIEW
CAPT(S) Andrew A. King, USN, Branch Head, Expeditionary Warfare, Mobility, and Sustainment, Assessment Division, Office of the Deputy Chief of Naval Operations for Resources, Requirements, and Assessments, N812C

0915 U.S. NAVY SEA BASING ANALYSES—FORCE CLOSURE
LCDR Christofer M. Collins, USN, Expeditionary Warfare Analyst, Assessment Division, Office of the Deputy Chief of Naval Operations for Resources, Requirements, and Assessments, N812C3

0945 U.S. NAVY SEA BASING ANALYSES—FORCE ASSEMBLY
LCDR Christofer M. Collins, USN, Expeditionary Warfare Analyst, Assessment Division, Office of the Deputy Chief of Naval Operations for Resources, Requirements, and Assessments, N812C3

1030 U.S. NAVY SEA BASING ANALYSES—AIR EMPLOYMENT
CDR(S) Brian A. Hoyt, USN, Expeditionary Warfare Analyst, Assessment Division, Office of the Deputy Chief of Naval Operations for Resources, Requirements, and Assessment, N812C2

1200 U.S. NAVY SEA BASING ANALYSES—SURFACE EMPLOYMENT
LCDR Christofer M. Collins, USN, Expeditionary Warfare Analyst, Assessment Division, Office of the Deputy Chief of Naval Operations for Resources, Requirements, and Assessment, N812C3

1245 U.S. NAVY SEA BASING ANALYSES—MARITIME PREPOSITIONED FORCE (FUTURE) [MPF(F)] FORCE
LCDR Eric R. Horning, USN, Expeditionary Warfare Analyst, Assessment Division, Office of the Deputy Chief of Naval Operations for Resources, Requirements, and Assessment, N812E

1315 U.S. NAVY SEA BASING ANALYSES—MPF(F) SURVIVABILITY
CDR(S) Brian A. Hoyt, USN, Expeditionary Warfare Analyst, Assessment Division, Office of the Deputy Chief of Naval Operations for Resources, Requirements, and Assessment, N812C2

1400 U.S. NAVY SEA BASING ANALYSES—SUSTAINMENT
CDR John V. Harmon, USN, Sustainment Analyst, Assessment Division, Office of the Deputy Chief of Naval Operations for Resources, Requirements, and Assessment, N812C1

1430 HEAVY-LIFT AIRCRAFT—REPORT FROM THE JOINT VERTICAL AIRCRAFT TASK FORCE (JVATF)
Michael J. Walsh, Staff Specialist, Office of the Director for Defense Systems, Office of the Under Secretary of Defense for Acquisition, Technology, and Logistics/Land Warfare and Munitions, OUSD/AT&L/LW&M

Closed Session: Committee Members and NRC Staff Only

1530 COMMITTEE DISCUSSION—RECAP OF DAY 2
Moderators:
Harry W. Jenkins, Jr., Committee Co-Chair
Richard L. Wade, Committee Co-Chair

1830 END SESSION

Thursday, September 9, 2004

Closed Session: Committee Members and NRC Staff Only

0830 CONVENE—COMMITTEE DISCUSSION
Harry W. Jenkins, Jr., Committee Co-Chair
Richard L. Wade, Committee Co-Chair
Arul Mozhi, Senior Program Officer, Naval Studies Board

Data-Gathering Meeting Not Open to the Public: Classified Discussion (Secret)

0900 U.S. MARINE CORPS SEA BASING ANALYSES—OVERVIEW
James Strock, Deputy Director, Expeditionary Force Development Center, Marine Corps Combat Development Command

0915 U.S. MARINE CORPS SEA BASING ANALYSES—MPF(F) REQUIREMENTS
William A. Sawyers, Head, Mission Area Analysis Branch, Studies and Analysis Division, Marine Corps Combat Development Command
Daniel Purcell, Analyst, Mission Area Analysis Branch, Studies and Analysis Division, Marine Corps Combat Development Command

1045 ANALYSES OF SEA BASING CONNECTORS
Maj Scott Kish, USMC, Analyst, Sea Base Branch, Office of the Deputy Chief of Naval Operations for Warfare Requirements and Programs, N703M

1230 RESEARCH AND DEVELOPMENT EFFORTS FOR MPF(F)
Arthur Rausch, Head, Expeditionary/Logistics/U.S. Marine Corps Systems Division, Naval Surface Warfare Center, Carderock Division

1330 INDUSTRY PERSPECTIVE ON SEA BASING TECHNOLOGIES
Mark Rice, Principal Engineer, Maritime Applied Physics Corporation
William Peterson, Naval Architect, Maritime Applied Physics Corporation

1445 SHIPBUILDER'S PERSPECTIVE ON SEA BASING TECHNOLOGIES
Matthew P. Tedesco, National Steel and Shipbuilding Company (NASSCO) Representative to the National Shipbuilding Research Program (NSRP) Industry Design Review Board for Office of Naval Research

1545 INDUSTRY PERSPECTIVE ON AFLOAT SEA BASING CONCEPTS
Louis M. Lambremont, Vice President, Business Development, MAERSK Line, Limited

Closed Session: Committee Members and NRC Staff Only

1645 COMMITTEE DISCUSSION—RECAP OF DAY 3
Moderators:
Harry W. Jenkins, Jr., Committee Co-Chair
Richard L. Wade, Committee Co-Chair

1730 END SESSION

Friday, September 10, 2004

Closed Session: Committee Members and NRC Staff Only

0830 CONVENE—COMMITTEE DISCUSSION
Harry W. Jenkins, Jr., Committee Co-Chair
Richard L. Wade, Committee Co-Chair
Arul Mozhi, Senior Program Officer, Naval Studies Board

Data-Gathering Meeting Not Open to the Public:
Classified Discussion (Secret)

0900 SEA BASING TECHNOLOGICAL AND OPERATIONAL CHALLENGES
BrigGen Robert E. Schmidle, USMC, Director, Expeditionary Force Development Center, Marine Corps Combat Development Command

Closed Session: Committee Members and NRC Staff Only

1015 REPORT DISCUSSION AND CONSENSUS ON DRAFT CONCLUSIONS/RECOMMENDATIONS
Moderators:
Harry W. Jenkins, Jr., Committee Co-Chair
Richard L. Wade, Committee Co-Chair

1230 CONSENSUS ON DRAFT CONCLUSIONS/RECOMMENDATIONS; DEVELOP CHAPTER ANNOTATED OUTLINES
Moderators:
Harry W. Jenkins, Jr., Committee Co-Chair
Richard L. Wade, Committee Co-Chair

1600 ADJOURN

C

Acronyms and Abbreviations

AB	advanced base
AFSB	Afloat Forward Staging Base
AMC	Air Mobility Command
AMC-X	Advanced Mobility Concept (aircraft)
ARF	Army Regional Flotilla
ARG	Amphibious Ready Group
ATD	Advanced Technology Demonstration
BCT	Brigade Combat Team
B-HSC	beachable high-speed connector
B-HSV	beachable high-speed vessel
BWB	blended wing body (flying wing)
C2	command and control
C3	command, control, and communications
C4ISR	command, control, communications, computers, intelligence, surveillance, and reconnaissance
CLF	combat logistics force
CMC	Commandant, Marine Corps
CNO	Chief of Naval Operations
COCOM	combatant commander
COD	carrier onboard delivery
CONOPS	concept of operations

CONUS	continental United States
CSG	Carrier Strike Group
CVBG	Carrier Battle Group
DLA	Defense Logistics Agency
DOD	Department of Defense
DOF	degree of freedom
DSB	Defense Science Board
ECC	efficient combined container
EFV	expeditionary fighting vehicle
ESG	Expeditionary Strike Group
FCS	Future Combat System
FOC	full operational capability
FPR	fan pressure ratio
FY	Fiscal Year
HLLCAC	heavy-lift landing craft, air cushion
HLS	heavy-lift ship
HLVTOL	heavy-lift vertical takeoff and landing
HSC	high-speed connector
HSS	high-speed sealift
HSS-SD	high-speed sealift–shallow draft
HSV	high-speed vessel
HULA	ultra heavylift aircraft
ILP	integrated landing platform
IOC	initial operational capability
ISB	intermediate support base
ISO	International Organization for Standardization
ISR	intelligence, surveillance, and reconnaissance
JCIDS	Joint Capabilities Integration and Development System
JFCOM	Joint Forces Command
JLOTS	Joint Logistics Over the Shore (program)
JMIC	Joint Modular Inter-Modal Container (program)
JSF	Joint Strike Fighter
JTF	Joint Task Force
LAV	light assault vehicle
LCAC	landing craft, air cushion
LCAC-X	landing craft, air cushion-experimental

LCM	landing craft, mechanized
LCS	landing craft, support
LCU	landing craft, utility
LCU(R)	landing craft, utility (replacement)
L/D	lift to drag
LHA	amphibious assault ship, general purpose
LHD	amphibious assault ship, multipurpose
LMSR	large, medium-speed roll on/roll off
LO/LO	lift on/lift off
LPD	amphibious assault transport, dock
LSD	landing ship, dock
MEB	Marine Expeditionary Brigade
MEU	Marine Expeditionary Unit
MLP	mobile landing platform
MPF	Maritime Prepositioning Force
MPF(A)	Maritime Prepositioning Force (Aviation)
MPF(F)	Maritime Prepositioning Force (Future)
MPSRON(F)	Maritime Prepositioning Squadron (Future)
MSC	Military Sealift Command
M-X	Materiel, Experimental
NASA	National Aeronautics and Space Administration
NAVAIR	Naval Air Systems Command
nmi	nautical mile
NOC	Naval Operating Concept
NRC	National Research Council
OMFTS	Operational Maneuver From the Sea
ONR	Office of Naval Research
OSD	Office of the Secretary of Defense
PASCAT	partial air cushion supported catamaran
R&D	research and development
RO/RO	roll on/roll off
S&T	science and technology
SCN	Ship Construction, Navy
SDHSS	shallow draft high-speed ship
SHP	shaft horsepower
SOCOM	Special Operations Command

SOF	Special Operations Forces
SSTOL	super short takeoff and landing
STOL	short takeoff and landing
STOM	Ship-to-Objective Maneuver
STOVL	short takeoff and vertical landing
TEU	20-ft-equivalent unit
TRANSCOM	Transportation Command
TRL	Technology Readiness Level
TSV	theater support vessel
USN	U.S. Navy
USNS	U.S. naval ship
VTOL	vertical takeoff and landing